# Mathematics for Industry

## Volume 36

**Aims & Scope**

The meaning of "Mathematics for Industry" (sometimes abbreviated as MI or MfI) is different from that of "Mathematics in Industry" (or of "Industrial Mathematics"). The latter is restrictive: it tends to be identified with the actual mathematics that specifically arises in the daily management and operation of manufacturing. The former, however, denotes a new research field in mathematics that may serve as a foundation for creating future technologies. This concept was born from the integration and reorganization of pure and applied mathematics in the present day into a fluid and versatile form capable of stimulating awareness of the importance of mathematics in industry, as well as responding to the needs of industrial technologies. The history of this integration and reorganization indicates that this basic idea will someday find increasing utility. Mathematics can be a key technology in modern society.

The series aims to promote this trend by 1) providing comprehensive content on applications of mathematics, especially to industry technologies via various types of scientific research, 2) introducing basic, useful, necessary and crucial knowledge for several applications through concrete subjects, and 3) introducing new research results and developments for applications of mathematics in the real world. These points may provide the basis for opening a new mathematics-oriented technological world and even new research fields of mathematics.

To submit a proposal or request further information, please use the PDF Proposal Form or contact directly: Swati Meherishi, Executive Editor (swati.meherishi@springer.com).



More information about this series at https://link.springer.com/bookseries/13254

Robert McKibbin · Graeme Wake · Osamu Saeki
Editors

# Proceedings of the Forum "Math-for-Industry" 2019

## Mathematics for the Primary Industries and the Environment

Springer

*Editors*
Robert McKibbin
Massey University
Auckland, New Zealand

Osamu Saeki
Kyushu University
Fukuoka, Japan

Graeme Wake
School of Mathematical and Computational Sciences
Massey University
Auckland, New Zealand

ISSN 2198-350X ISSN 2198-3518 (electronic)
Mathematics for Industry
ISBN 978-981-19-1153-8 ISBN 978-981-19-1154-5 (eBook)
https://doi.org/10.1007/978-981-19-1154-5



This Springer imprint is published by the registered company Springer Nature Singapore Pte Ltd.
The registered company address is: 152 Beach Road, #21-01/04 Gateway East, Singapore 189721, Singapore

# Organization

**Forum "Math-for-Industry" 2019 -Mathematics for the Primary Industries and the Environment-**

Date: November 18-21, 2019
Venue: Massey University, Albany campus North Shore, Auckland New Zealand

- start 0930, finish 4:30 each day
17/11 Sunday afternoon/evening from 4 pm: IMI-IAB, APCMfI (Council and AGM), Editorial Board meetings
18/11 Late Monday afternoon: welcome party
20/11 Wednesday afternoon: excursion (there will be a fee for those who participate in this)

**Organising Committee**

Robert McKibbin [Chair] (Massey University, NZ)
Luke Fullard (Massey University, NZ)
Winston Sweatman (Massey University, NZ)
Mark McGuinness (Victoria University, NZ)
Kenji Kajiwara (Kyushu University, Japan)
Osamu Saeki (Kyushu University, Japan)
Masato Wakayama (Kyushu University, Japan)
Yasuhide Fukumoto (Kyushu University, Japan)

**FMfI2019 will include:**

* Invited talks
* Emerging Researcher talks
* Student posters
* Industry presentations + mix-and-mingle with reps
* Excursion + banquet NZ-style
* Meetings of APCMfI + IMI-IAB + Journal Editorial Boards

**Invited Speakers Committee**

Graeme Wake [Chair] (Massey University, NZ)
Graham Weir (Massey University, NZ)
Gaven Martin (Massey University, NZ)
Alona Ben-tal (Massey University, NZ)
Troy Farrell (Queensland University of Technology, Australia)
Masato Wakayama (Kyushu University, Japan)
Yasuhide Fukumoto (Kyushu University, Japan)
Jin Cheng (Fudan University, PR China)

**Accommodation:**

* On-campus accommodation in the student village (2 minutes walk)
* Hotels and motels nearby (some within 10 minutes walk)
* Plen y of hotels and motels in Auckland City (20 mins by car or bus)
* estaurants/food hall on campus; large shopping centre nearby with food hall and other cafes, restaurants

**Theme may include:**

*** efficiency and effectiveness of production, processing and transportation within primary industries**
For example:

- Farm systems management
- Dairy, meat, agricultural and horticultural food products
- Wine and fruit production
- Fisheries
- Fishing in NZs Exclusive Economic Zone from 26 degrees south.
- Forestry
- Mining, and steel production Electrical power production (including geothermal)

*** environmental effects/consequences/problems** For example:

- Water Quality
- Air pollution
- Plastics pollution
- Pest eradication [NZ has an aim to be predator-free by 2050]
- Climate change
- Consequences of increased intensive farming practices on the environment
- Energy generation and use (of any kind)

*** anything related to these, including biological, sociological, economics, quantitative methods, etc**

Massey University
COLLEGE OF SCIENCES

Institute of Mathematics for Industry
Kyushu University

Asia Pacific Consortium of Mathematics for Industry

# Call for Posters

## Forum "Math-for-Industry 2019"

Mathematics for the Primary Industries and the Environment

## Poster Session

## November 19, 2019

## Massey University, New Zealand

FMfI2019 will have a poster session for graduate students and postdocs. Poster presentations in any area of Mathematics (not necessarily related to the theme of the Forum directry) are invited.

Please send the title to "fmfi@imi.kyushu-u.ac.jp"

by October 31th, 2019.

For more details, please visit :

http://apcmfi.org/fmfi2019/index.html

# FMfI 2019 Mathematics for the Primary Industries and the Environment: Programme

**Sunday 17 November 2019** — Round Room/Functions Room

| | |
|---|---|
| 2.00–6.00 | Registration |
| 2.00 | IMI IAB |
| 3.00 | APCMfI Board |
| 4.00 | APCMfI AGM |
| 5.00 | Journal Boards |

**Monday 18 November 2019** — Round Room

| | |
|---|---|
| 08.30–09:30 | Registration |
| 09.30–10.00 | Welcome and Opening |
| 10.00–10.30 | Refreshment break |
| 10.30–11.10 | Toshihiko Takemura<br>Japan |
| 11.10–11.50 | Mike Plank<br>NZ |
| 11.50-12.30 | Hyun Min Kim<br>Korea |
| 12:30-1:30 | Lunch |
| 1.30–2.10 | Elliot Carr<br>Australia |
| 2.10–2.50 | Tammy Lynch<br>NZ |
| 2.50–3.10 | Break |
| 3.10–3.45 | Tanya Soboleva<br>MPI, NZ |
| 3.45–4.20 | Conrad Edwards<br>Transpower, NZ |
| 4:20–4.55 | Lisa Thomasen<br>Fonterra, NZ |
| 4.55–5.30 | Tony Pleasants<br>Al Rae Centre, NZ |
| 5.30-6.30 | Welcome Party |

**Tuesday 19 November 2019** — Round Room

| | |
|---|---|
| 09.30–10.10 | Lu Wenlian<br>PRC |
| 10.10–10.50 | Steven Psaltis<br>Australia |
| 10.50–11.10 | Break |
| 11.10–11.50 | Brent Clothier<br>NZ |
| 11.50-12.30 | Takasi Okayasu<br>Japan |
| 12:30-1:30 | Lunch |
| 1.30–2.10 | Graeme Hocking<br>Australia |
| 2.10–2.50 | Nicholas Long<br>NZ |
| 2.50–3.10 | Break |
| 3.10–5.30 | Poster Session |

**Wednesday 20 November 2019** — Round Room

| | |
|---|---|
| 09.30–10.10 | Nurul Syaza Abdul Latif<br>Malaysia |
| 10.10–10.50 | Yongwimon Lenbury<br>Thailand |
| 10.50–11.10 | Break |
| 11.10–11.50 | Aroon Parshotam<br>NZ |
| 11.50-12.30 | Brendan Florio<br>Australia |
| 12:30-1:30 | Lunch<br>Discussion/Workshop** |
| 1.30–6:00 | Free time<br>Self-propelled exploration of Downtown Auckland |
| 6.00–9.00 | Dinner Cruise |

**Thursday 21 November 2019** — Round Room

| | |
|---|---|
| 09.30–10.10 | Eloise Tredenick<br>Australia |
| 10.10–10.50 | Akiko Satake<br>Japan |
| 10.50–11.10 | Break |
| 11.10–11.50 | Mark McGuinness<br>NZ |
| 11.50-12.30 | John Burnell<br>NZ |
| 12:30-1:30 | Lunch |
| 1.30–2.10 | Mark Nelson<br>Australia |
| 2.10–2.50 | Kaji Shizuo<br>Japan |
| 2.50–3.10 | Closing |
| 3.10–3.30 | Refreshments |

**Session Chairs**

| | |
|---|---|
| Monday 10.30–12.30 | Winston Sweatman |
| 1.30–2.50 | Mike Plank |
| 3.10–5.30 | Graeme Wake |
| Tuesday 09.30–10.50 | Mark McGuinness |
| 11.10–12.30 | Phil Broadbridge |
| 1.30– 2.50 | Graham Weir |
| 3.10–5.30 | Robert McKibbin |
| Wednesday 09.30–10.50 | Graeme Wake |
| 11.10–12.30 | Graeme Hocking |
| Thursday 09.30–10.50 | Tammy Lynch |
| 11.10–12.30 | Mark Nelson |
| 1.30–2.50 | Luke Fullard |

The Round Room and Functions Room are in the Atrium Building

** Mathematical modelling: what could it look like in the classroom?
(Leader: Kerri Spooner)

# Preface

The use of mathematics in primary industries (such as agriculture, horticulture, forestry, food production) as well as in the environmental and medical areas has accelerated in many countries over the last decade. Methods used to extend right across the spectrum of the mathematical sciences, including complex systems, stochastic processes, and numerical techniques. Non-mathematicians are increasingly realising that they need access to, and guidance in, the use of these methods to underpin their decision support processes.

Those of us who work in this area need to be aware that in many countries there is a long-held apprehension towards mathematics that is not always well-understood and addressed by us. So, it requires a developed and patient attitude in the delivery of advice.

Massey University has long been an advocate of the importance of how mathematics and statistics can be useful in industry, and this continues with strong support for, and participation in Mathematics-in-Industry study groups held in New Zealand and Australia, and elsewhere around the world. Applications in the primary industries and the environment are important to our country and, indeed, everywhere.

It is with all this in mind that we organised the conference: Forum "Math-for-Industry" 2019 (FMfI2019) for which the theme was "Mathematics in the Primary Industries and the Environment". It was held on the Auckland campus of Massey University in New Zealand during 17–21 November 2019. New Zealand is the southernmost member country within the Asia-Pacific Consortium of Mathematics-for-Industry (APCMfI). We were driven by the reality that for our host country these industries are the largest and underpin the economy. The topics covered in the Forum included contributions related to the dairy industry, farming and horticultural systems and their management, energy production, infrastructure, fisheries and forestry, water supplies, air pollution, and optimisation techniques.

Several New Zealand industries presented a local perspective on industrial and environmental matters of importance (NZ Ministry for Primary Industries' NZ Food Safety, Transpower NZ, Fonterra NZ's Co-operative Group Ltd, and Massey University's Al Rae Centre for Breeding and Genetics).

This volume is a good and timely indication of the growth of these applications in our region, although it contains only a small sample of papers on the topics of the Forum. A healthy number of graduate students attended, with several winning awards provided by the Institute of Mathematics-for-Industry (IMI, Kyushu University) for the high quality of their posters and presentations. The students who presented posters and gave talks about their work are the future leaders in APCMfI. They are valuable members of the "Math-for-Industry" community, and were particularly welcome at the Forum. This has shown that the widening scope for a large variety of applications to attract our mathematical talent and that many techniques are quite ubiquitous in nature.

We would like to thank the participants of the Forum and members of the Invited Speakers Committee. These, the invited speakers, and the conference administrators, especially Sue di Leo and Annette Warbrooke from Massey, and Seiko Sasaguri from IMI, ensured that the event went very smoothly. The meeting was a worthy contribution to its cause: to highlight the growing diversity and value of quantitative methods to address our problems and challenges.

Auckland, New Zealand
November 2021

Robert McKibbin
Graeme Wake
On behalf of the Editorial Board of the
Proceedings of FMfI2019

# Contents

# About the Editors

**Robert McKibbin** now retired from New Zealand's Massey University, is Emeritus Professor of Applied Mathematics. An alumnus of both the University of Canterbury (M.Sc. in Mathematics), and the University of Auckland (Ph.D. in Engineering Science), he is Fellow of the NZ Mathematical Society. He was awarded the NZ Geophysics Prize 1982, the NZ Royal Society's Hamilton Memorial Prize 1984, and the ANZIAM Medal 2012. He has been involved with Industrial Mathematics within NZ and also with the Asia Pacific Consortium (APCMfI) initiated by Kyushu University's Institute of Mathematics for Industry; he was Member of the latter's International Advisory Board for several years. His research interests are in mathematical modeling of heat and mass transfer processes and particularly those of geophysical phenomena associated with geothermal systems, including thermally driven underground fluid convection and atmospheric particle transport associated with volcanic and other eruptive mechanisms.

**Graeme Wake** is now Adjunct Professor of two Auckland University campuses in New Zealand: Auckland University of Technology and Massey University (also is Emeritus Professor of the latter). He has served in four different University campuses in New Zealand, was Postdoctoral Scholar (1970), and much later Visiting Fellow at three different Oxford University Colleges. He was awarded the ANZIAM Medal for services to, and excellence in, ANZ Applied Mathematics in 2006, a Doctor of Science degree from Victoria University of Wellington in 1997, and a Fellowship of the Royal Society of New Zealand in 2004. He held two Visiting Fulbright awards, one (incoming) from the USA) and one as Visiting Fulbright Professor to USA. In 2015, he was Foundation Leader of the Group "Mathematics-in-Industry for New Zealand (MINZ)" which organizes Study Groups like those that exist now in many countries, and which started in Oxford in 1968. Graeme was the initial NZ representative of the Asia Pacific Consortium when it was formed in 2015. His research encompasses industrial mathematics, mathematical biology, applied differential, and functional differential equations.

**Osamu Saeki** is Distinguished Professor at Kyushu University. He received his Ph.D. in Mathematics from the University of Tokyo. He was awarded the Takebe Katahiro Prize 1996 and the Geometry Prize 2015 from the Mathematical Society of Japan. He was involved in establishing the Institute of Mathematics for Industry (IMI), Kyushu University, launched in 2011. He is also engaged in the education of industrial mathematics and is Coordinator of the WISE program "Graduate Program of Mathematics for Innovation", supported by MEXT, Japan. His current research interests concern topology, singularity theory, topology of low-dimensional manifolds, knot theory, and visualization of large scale data.

# A Model for Coating Steel—Draining Under Gravity

**Graeme C. Hocking**

## 1 Introduction

The process of galvanization of steel has been considered at several Maths-in-Industry study groups to try to identify the reason for flaws in the coating (see Hocking et al. 2010, 2016). The flaws included pitting on the surface and droplet formation (which occur during thinner coatings and faster travel times) (Hocking et al. 2010) and formation of "Bananas" of very poor quality coating that grow in from the sides (for lower pressures from the air knife and for wider sheets) (Hocking et al. 2016).

In the full industrial process, the sheet is first coated with aluminium, then heated to a high temperature, before being passed through a bath of molten alloy at about 1–2 m/s. As the steel is drawn upward, a high velocity air jet, called an air knife, blows off excess liquid which drains back into the bath (see Fig. 1). After this "stripping" of excess fluid the sheet rises and cools in the air before a final stage of water cooling. Baffles are sometimes employed at the sides of the sheet to direct the air from the air knife, although in theory these would remove the two-dimensionality of the flow and cause some inconsistencies near the edges, unless very carefully deployed. Tuck (1983) and Hocking et al. (2010, 2011) derived and analysed the equations for the problem of jet-stripping, and the results are summarized briefly in Sect. 3.

In this article, however, we consider the problem of draining under gravity, with no air jet, making the assumption of unidirectional flow (Batchelor 1967). Some new solutions are presented using perturbation techniques, and the results are discussed.

G. C. Hocking (✉)
Mathematics and Statistics, Murdoch University, Murdoch, WA 6150, Australia
e-mail: G.Hocking@murdoch.edu.au

 
R. McKibbin et al. (eds.), *Proceedings of the Forum "Math-for-Industry" 2019*,
Mathematics for Industry 36, https://doi.org/10.1007/978-981-19-1154-5_1

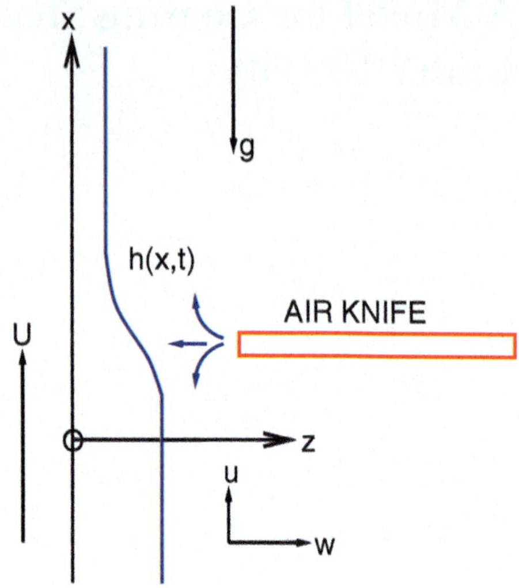


**Fig. 1** Sketch of the air knife and coating configuration

## 2 Unidirectional Flow—Gravity Draining

In the case of gravity draining and in the absence of the air knife, we can take the full equations of a Newtonian fluid and assume a steady flow of a viscous, incompressible fluid. Defining a coordinate system $(x, y, z)$ with corresponding velocity components $(u, v, w)$ and such that $x$ is directed upward (see Fig. 1) and taking the flow to be unidirectional so that $v = w = 0$ the result is a balance between the viscous terms, gravity, and pressure.

Starting with the continuity equation, we find, noting that derivatives with respect to $y$ and $z$ are zero,

$$\frac{\partial u}{\partial x} + \frac{\partial v}{\partial y} + \frac{\partial w}{\partial z} = 0 \text{ so that } \frac{\partial u}{\partial x} = 0, \tag{1}$$

which means that $u(y, z)$ is a function of $y$ and $z$ only. If this outcome is included, the momentum equations

$$\frac{\partial u}{\partial t} + u\frac{\partial u}{\partial x} + v\frac{\partial u}{\partial y} + w\frac{\partial u}{\partial z} = -\frac{1}{\rho}\frac{\partial p}{\partial x} - g + \nu(u_{xx} + u_{yy} + u_{zz}) \tag{2}$$

$$\frac{\partial v}{\partial t} + u\frac{\partial v}{\partial x} + v\frac{\partial v}{\partial y} + w\frac{\partial v}{\partial z} = -\frac{1}{\rho}\frac{\partial p}{\partial y} + \nu\nabla^2 v \tag{3}$$

$$\frac{\partial w}{\partial t} + u\frac{\partial w}{\partial x} + v\frac{\partial w}{\partial y} + w\frac{\partial w}{\partial z} = -\frac{1}{\rho}\frac{\partial p}{\partial z} + \nu\nabla^2 w \tag{4}$$

indicate that $p_y = p_z = 0$ from (3, 4), leaving that $p(x)$ is a function of $x$ only. In the absence of the air jet, and noting the pressure on the surface must match atmospheric pressure, then $p'(x) = 0$. Omitting all terms that we have deemed to be negligible, we are left with a Poisson equation

$$u_{yy} + u_{zz} = g/\nu. \tag{5}$$

## 2.1 Broad Flat Sheet

Assuming that the sheet of metal is broad and flat, then the flow is two dimensional, in which case $u(y, z)$ becomes a function of $z$ only (Tuck 1983). The boundary conditions under gravitational draining are that at the solid-liquid interface, the fluid sticks, while on the outer edge of the coating, $z = h$, a constant, there must be no shear stress. Therefore the problem for a broad, flat sheet is

$$\frac{d^2u}{dz^2} = g/\nu, \text{ with}$$
$$u = U, \ z = 0,$$
$$u_z = 0, \ z = h.$$

The solution to this can be obtained by integration, giving

$$u(z) = U + \frac{g}{2\nu}(z - 2h)z,$$

but this is NOT the complete solution because we do not know the value of the coating thickness, $h$, even though we have the velocity profile. The correct extra condition, as discussed by earlier researchers (Elsaadawy et al. 2007; Thornton and Graff 1976; Tuck 1983) is that the amount of fluid carried upward is the maximum that can be sustained by the sheet, that is that the upward flux of coating material is maximized. Noting that the flux, $Q$, is

$$Q = \int_0^h u(z)dz = -\frac{g}{3\nu}h^3 + Uh,$$

we can obtain the maximum by finding the value of $h$ that maximizes $Q$, i.e. where $\frac{dQ}{dh} = 0$, leading to

$$-\frac{g}{\nu}h^2 + U = 0 \tag{6}$$

$$\text{and so } h^* = \sqrt{\frac{U\nu}{g}}, \tag{7}$$

and therefore the speed on the boundary is $u(h^*) = U/2$, while the maximum flux is $Q_{max} = \frac{2}{3}U^{2/3}(\nu/g)^{1/2}$.

## 2.2 Circular Object

Our main interest in this article is the gravitational flow (in the absence of the air knife) for different shapes. Therefore, to continue we consider an axisymmetric flow situation. In cylindrical, polar coordinates, following the same process as above, the problem is

$$\frac{1}{r}\frac{\partial}{\partial r}\left(r\frac{\partial u}{\partial r}\right) = \frac{g}{\nu}$$
$$\text{subject to } u = u_b \text{ on } r = R$$
$$u'(r) = 0 \text{ on } r = h$$

and integrating we find the velocity to be

$$\begin{aligned} u(r) &= \frac{gr^2}{4\nu} + A\log r + B \\ &= u_b + \frac{g}{4\nu}\left(r^2 - R^2\right) + \frac{gh^2}{2\nu}\log(R/r). \end{aligned}$$

As before, $h$ is unknown and constant, but the upward flux of the coating is given by

$$\begin{aligned} Q/2\pi &= \int_R^h u(r) r\, dr \\ &= u_b\frac{(h^2 - R^2)}{2} + \frac{g}{4\nu}\left(\frac{(h^4 - R^4)}{4} - \frac{(h^2 - R^2)R^2}{2}\right) \\ &\quad + \frac{gh^2}{8\nu}\left((h^2 - R^2) + 2h^2\log(R/h)\right) \end{aligned}$$

and as before we can find the maximum flux; in this case

$$\frac{dQ}{dh} = 2\pi h\left(u_b + \frac{g}{4\nu}(h^2 - R^2) + \frac{g}{4\nu}\left(h^2 - R^2 - -4^2\log(h/R)\right)\right) = 0$$

so that

$$(h^2 - R^2) - 2h^2\log(h/R) + \frac{2\nu u_b}{g} = 0. \tag{8}$$

At the outer edge of the coating

$$u(h) = u_b + \frac{g}{4\nu}\left(h^2 - R^2\right) - \frac{gh^2}{2\nu}\log(h/R)$$
$$= u_b - u_b/2 \text{ so that } u(h^*) = \frac{1}{2}u_b.$$

This is the same relative velocity as for the broad sheet and so the question is whether this is the case for all shapes? In fact, Tuck et al. (1983) showed that this is indeed the case for any shape, and so the problem becomes much simpler in that circumstance.

## 3 "Mainly" Unidirectional Flow

The broad sheet solution in Sect. 2.1 can be improved for the inclusion of the air knife by allowing there to be a reasonably small variation from purely vertical flow, so that we assume the flow to be *mainly* unidirectional (Hocking et al. 2010; Tuck 1983). In that case, $w$ is no longer assumed to be zero, and the equations become

$$u_t + uu_x + wu_z = -\frac{1}{\rho}p_x + \frac{\mu}{\rho}(u_{xx} + u_{zz}) - g$$
$$w_t + uw_x + ww_z = -\frac{1}{\rho}p_z + \frac{\mu}{\rho}(w_{xx} + w_{zz})$$
$$u_x + w_z = 0$$

with boundary conditions;

$$u = U, \quad w = 0 \quad \text{on } z = 0,$$
$$\begin{cases} \mu u_z = \tau_a(x), \\ p - p_a(x) = -\gamma\kappa, \\ h_t + uh_x = w, \end{cases} \qquad \text{on } z = h(x,t)$$

where $\gamma$ is surface tension, $\kappa \sim h_{xx}$ is the curvature, and $p_a(x), \tau_a(x)$ are the air pressure and shear stress due to the air knife.

After some work (Hocking et al. 2010), this gives a first order, nonlinear advection equation for the coating thickness, $h(x,t)$,

$$h_t + c(h, p_a', \tau)h_x = A(h, p_a'', \tau'),$$

where $c(h, p_a', \tau) = 1 - (P^* p_a' + S)h^2 + h\tau$, and $A(h, p_a'', \tau') = \frac{1}{3}h^3 P^* p_a'' - \frac{1}{2}h^2\tau'$, with $S$ for gravity (Stokes No.), and $P^*$ is a reference pressure. From the theory of first-order partial differential equations we know that $c(h, p'_a, \tau)$ is the speed of propagation of any perturbation to the surface, while $A(h, p''_a, \tau')$ is the amplification.

An analysis of this equation is given in Tuck (1983) and it was also considered in different ways in the two study groups (Hocking et al. 2010, 2016), and in more detail in Hocking et al. (2011). The analysis included the computation of steady-state solutions and the method of characteristics was used to consider the behaviour of small perturbations in the surface. These perturbations were found to be marginally stable, which is to say that they neither grow nor shrink as the flow proceeds.

The most important of these results are the steady-state solutions calculated given different values of the important parameters, since they provide the baseline behaviour. The result is a cubic for $h(x)$ at each point;

$$Q = f(h, x) = h + \frac{h^2}{2}G(x) - \frac{h^3}{3}(S + P'(x)),$$

where $G(x) = \tau + \gamma'(x)$. Each different $x$ location has a slightly different cubic, and so gives a different maximum $Q$. The problem is therefore to find the minimum of the maximum fluxes, since that must be the maximum of the whole system. This process is described in Hocking et al. (2010), Tuck (1983) and it provides an accurate representation of coating under the air knife in the case of a broad sheet. The case of vertical characteristics occurs when $c(h, p'_a, \tau) = 0$, i.e.

$$1 - (P^* p'_a + S)h^2 + h\tau = 0$$

$$\text{leading to } h_V = \frac{\tau}{2P^* p_a}\left(1 \pm \sqrt{1 + \frac{4P^* p'_a}{\tau^2}}\right)$$

with the result that if $h > h_V$ at any point, disturbances propagate backwards, while if $h < h_V$ at any point, disturbances propagate forwards. This means that if a dip forms, depending on the exact parameter values, it may steepen at its leading edge or trailing edge causing a "wave breaking" phenomenon, leading to a small blob formation, but the depression will remain. Such blobs have been observed in practice.

## 4 More Solutions: Perturbations to a Flat Sheet

Returning to the draining flow problem that is the focus of this article, we can compute solutions for draining flows for variations on the flat sheet (Sect. 2.1) and a circular shape (Sect. 2.2). Tuck et al. (1983) developed an iterative numerical scheme, while (Howison and King 1989) used complex variable mappings to computing coatings for different shapes, and (Hocking 2020) used a fundamental singularity numerical method to compute the surface shape for elliptical objects. One solution found using the method of Tuck et al. (1983) is shown in Fig. 2. At the end of a broad sheet the

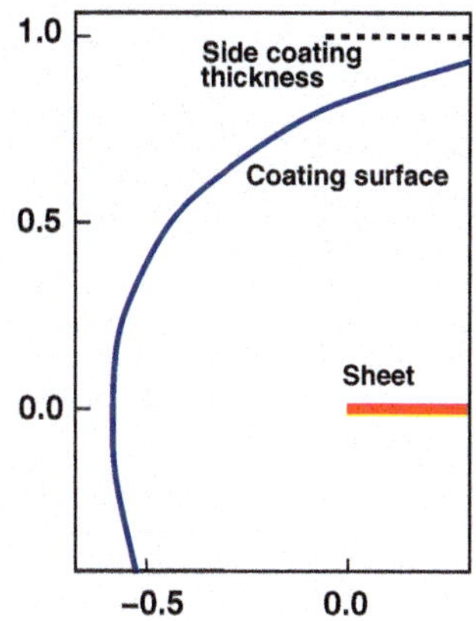


**Fig. 2** Computed solution from Tuck et al. (1983). The narrowest point is at the end, the thickness at that point being 63% of the maximum side coating thickness

coating is around 63% of the thickness along the side, thus making it more vulnerable to gaps in the coating if there are inconsistencies in the jet-stripping process.

A perturbation method to compute solutions to the draining problem is now described, where the coating thickness for small deviations from a flat sheet can be calculated. Non-dimensionalizing with respect to velocity, $U$, and length, $L = \sqrt{U\nu/g}$, the problem becomes

$$\frac{\partial^2 u}{\partial y^2} + \frac{\partial^2 u}{\partial z^2} = 1 \tag{9}$$

$$\text{subject to} \quad u = 1, \quad \text{on } y = z_S, \tag{10}$$

$$\frac{\partial u}{\partial n} = 0, u = 1/2 \text{ on } y = z_F, \tag{11}$$

where $z_F$ is the as-yet unknown function of the coating shape and $z_S$ is the equation of the surface of the substrate

$$z_S = \varepsilon f_S(y), \tag{12}$$

where $\varepsilon$ is some small parameter that measures the height of the deviation relative to the thickness of the coating. Note that in this non-dimensional coordinate system the coating thickness for a perfectly flat sheet has unit value.

Now we seek a solution using perturbation techniques, by looking for a solution as an expansion in powers of $\varepsilon$, that is

$$u = u_0 + \varepsilon u_1 + \varepsilon u_2 + \cdots \tag{13}$$

$$z_F = 1 + \varepsilon f_1(y) + \varepsilon f_2(y) + \cdots, \tag{14}$$

where $f_j(y)$, $j = 1, 2, \ldots$ are successive approximations to the coating surface. Now since the small parameter appears implicitly in the functions we can follow (Van Dyke 1975) and expand $u$ about both $z = 0$ and $z = 1$, so on the substrate surface

$$u(y, z_S) = u_0(y,0) + \varepsilon f_S(y) u_{0z}(y,0) + \frac{1}{2}(\varepsilon f_S(y))^2 u_{0zz}(y,0) + \cdots$$
$$+ \varepsilon u_1(y,0) + \varepsilon^2 f_S(y) u_{1z}(y,0) + \cdots + \varepsilon^2 u_2(y,0) + \cdots = 1 \quad (15)$$

and on the surface of the coating

$$u(y, z_F) = u_0(y,1) + (\varepsilon f_1(y) + \varepsilon^2 f_2(y) + \cdots) u_{0z}(y,1)$$
$$+ \frac{1}{2}(\varepsilon f_1(y) + \varepsilon^2 f_2(y) + \cdots)^2 u_{0zz}(y,1) + \cdots$$
$$+ \varepsilon u_1(y,1) + \varepsilon(\varepsilon f_1(y) + \varepsilon^2 f_2(y) + \cdots) u_{1z}(y,1) + \cdots$$
$$+ \varepsilon^2 u_2(y,1) + \cdots = \frac{1}{2}, \quad (16)$$

where the subscript $z$ refers to the partial derivative. Similar expansions can be computed for all of the partial derivatives involved. The normal to the free boundary surface can be obtained as $\mathbf{n} = (-\varepsilon f_1'(y) - \varepsilon f_2'(y) \ldots, 1)$ so that the expansion of the zero surface stress equation $\nabla u \cdot n = 0$, evaluated on $z = 1$ is

$$u_{0z}(y,1) + (\varepsilon f_1(y) + \varepsilon^2 f_2(y) + \cdots) u_{0zz}(y,1) + (\varepsilon^2 f_1(y)^2 + \cdots) u_{0zzz}(y,1) + \cdots$$
$$+ \varepsilon u_{1z}(y,1) + \varepsilon(\varepsilon f_1(y) + \cdots) u_{1zz}(y,1) + \cdots + \varepsilon^2 u_{2z}(y,1) + \cdots$$
$$- \left(\varepsilon f_1'(y) + \varepsilon^2 f_2'(y) + \cdots\right)(u_{0y}(y,1) + (\varepsilon f_1(y) + \cdots) u_{0zy}(y,1) + \cdots$$
$$+ \varepsilon u_{1y}(y,1) + (\varepsilon f_1(y) + \cdots) u_{1zy}(y,1) + \cdots + \varepsilon^2 u_{2y}(y,1) + \cdots) = 0, \quad (17)$$

and so on. Thus we arrive at the leading order problem,

$$O(1) : \nabla^2 u_0 = 1 \text{ subject to}$$
$$u_0(y,0) = 1, \; u_0(y,1) = 1/2, \; \frac{\partial u_0}{\partial z}(y,1) = 0, \quad (18)$$

the original flat plate problem, for which the solution was $u_0 = \frac{1}{2}z^2 - z + 1$. The next level problem is

$$O(\varepsilon) : \nabla^2 u_1 = 0 \text{ subject to}$$
$$u_1(y,0) + f_S(y) u_{0z}(y,0) = 0 \text{ giving } u_1(y,0) = f_S(y), \quad (19)$$
$$u_1(y,1) + f_1(y) u_{0z}(y,1) = 0 \text{ giving } u_1(y,1) = 0, \quad (20)$$
$$u_{1z}(y,1) + f_1(y) u_{0zz}(y,1) - f_1'(y) u_{0y}(y,1) = 0. \quad (21)$$

Noting that $u_{0zz}(y,1) = 1$ and $u_{0y}(y,1) = 0$, the problem can be solved for $u_1(y,z)$ to satisfy (19) and (20), leaving the final condition (21) as

$$f_1(y) = -u_{1z}(y,1), \quad (22)$$

providing an estimate for the shape of the surface. A solution that satisfies Laplace's equation and (20) at $z = 1$ is

$$u_1(y, z) = \int_0^\infty A(k) \sinh k(z - 1) \cos ky \, dk. \tag{23}$$

We can solve for $u_1(y, 0) = f_S(y)$, providing us with $A(k)$, after which we can find $f_1(y)$ using (22).

## 4.1 An "Exact" Solution—Corrugation

A very simple solution can be obtained by noting that if we choose a single value of $k$ in Eq. (23), then we can make the perturbation to the surface $f_S(y) = \cos ky$, which represents a simple corrugated surface, and for which the solution can be written down as

$$u_1(y, z) = B \sinh k(z - 1) \cos ky \tag{24}$$

for any $k$, where $B$ is a constant. Substituting into (19),

$$u_1(y, 0) = -B \sinh k \cos ky = f_S(y) = \cos ky \text{ giving } B = -1/\sinh k,$$

and hence

$$u_1(y, z) = -\frac{\sinh k(z - 1)}{\sinh k} \cos ky \tag{25}$$

$$\text{so that } f_1(y) = \frac{k}{\sinh k} \cos ky, \tag{26}$$

and the shape of the surface coating is therefore

$$z_F = 1 + \varepsilon \frac{k}{\sinh k} \cos ky + O(\varepsilon^2) \tag{27}$$

when the substrate shape is $z_S = \varepsilon \cos ky$. As $k$ increases, the deviation of the surface reduces relative to the deviation on the surface of the sheet. Figure 3 shows a typical example with $k = 2.5$ and $\varepsilon = 0.1$. The dashed line indicates where a uniform coating would lie. The coating therefore smooths out the undulations. Thus the dips in the corrugation get a slightly thicker coating while the peaks get a thinner covering.

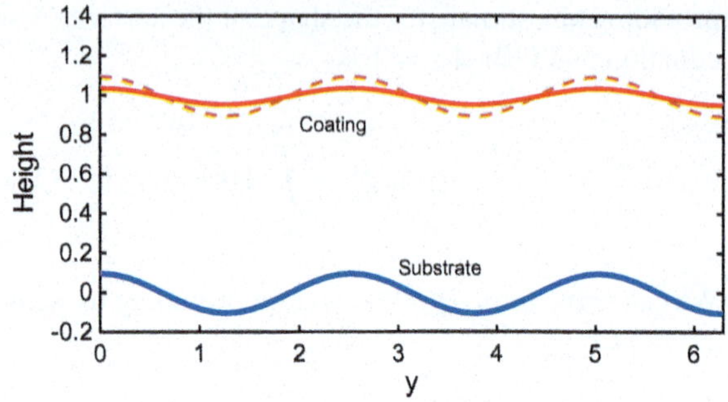


**Fig. 3** Surface shape for a corrugated perturbation with $k = 2.5$ and $\varepsilon = 0.1$. The dashed line represents where the surface coating would be if it had uniform thickness

## 4.2 *Single Bumps and Dips*

If we assume a Gaussian shape for the perturbation, then it is possible to compute the coating thickness for a single bump or dip. Therefore, we take the deviation to be of the form

$$f_S(y) = \pm\varepsilon e^{-y^2/\alpha}, \tag{28}$$

where the sign determines whether it is a dip or a bump. This time we can see that if $u_1(y, z)$ takes the form of (23) and

$$u_1(y, 0) = f_S(y) = e^{-y^2/\alpha} = \int_0^\infty A(k)\sinh(-k)\cos ky\, dk, \tag{29}$$

giving

$$A(k) = -2\left(\frac{\alpha}{\pi}\right)^{1/2} \frac{e^{-\alpha k^2}}{\sinh k} \tag{30}$$

using Fourier Cosine transforms, and hence

$$u_1(y, z) = -2\left(\frac{\alpha}{\pi}\right)^{1/2} \int_0^\infty \frac{e^{-\alpha k^2}\sinh k(z-1)}{\sinh k}\cos ky\, dk \tag{31}$$

and finally

$$f_1(y) = -u_{1z}(y, 1) = 2\left(\frac{\alpha}{\pi}\right)^{1/2} \int_0^\infty \frac{ke^{-\alpha k^2}}{\sinh k}\cos ky\, dk. \tag{32}$$

Figure 4 shows solutions for (a) a bump with $\varepsilon = 0.2$, and $\alpha = 1$ and (b) a dip of similar dimension. The bump leads to a coating of uniform thickness except for a slight thickening on either side of the peak. This is similar to the corrugated example in that the coating smooths over the flaw to some extent. However, the dip creates a

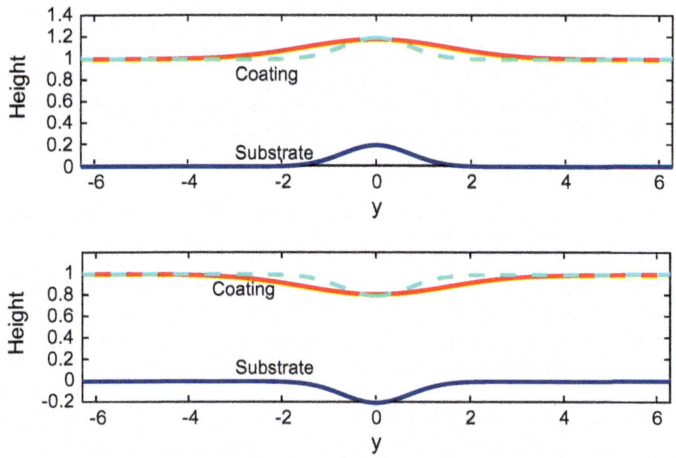


**Fig. 4** Surface shape for (**a**) a bump in the sheet, $f_S(y) = \exp(-y^2)$ and (**b**) a dip, $f_S(y) = -\exp(-y^2)$. The dashed line represents where the surface coating would be if it had uniform thickness

dip of similar depth in the coating, but it is quite a lot broader than the initial dip, resulting in a thinner coating on both sides of the dip. This is much more problematic once the coating is subject to an air jet, since this thinning may lead to gaps forming in the coating.

## 5 Final Remarks

Making the assumption of unidirectional flow, solutions can be obtained for the gravity draining of a coating of the alloy during the galvanization of steel. The resulting equations are relatively simple to solve numerically, but a simple perturbation analysis produces some of the important features of the coating thickness that evolve when small vertical bumps, dips, or corrugations form on the surface of the sheet.

These solutions under draining conditions are an important precursor to finding solutions when the air jet is also applied, since these deviations may explain some of the problems encountered in the manufacturing process.

## References

Batchelor GK (1967) An Introduction to fluid dynamics. Cambridge University Press, Cambridge. 0521098173

Elsaadawy EA, Hanumanth GS, Balthazaar AKS, McDermid JR, Hrymak AN, Forbes JF (2007) Coating weight model for the continuous hot-dip galvanizing process. Metall Mater Trans B 38B:413–424. https://doi.org/10.1007/s11663-007-9037-2

Hocking GC (2020) Draining under gravity in steel galvanization. ANZIAM J 61:C31–C44 (EMAC2019); Proceedings of 14th biennial engineering mathematics with application conference. https://doi.org/10.21914/anziamj.v61i0.15155

Hocking GC, Lavalle G, Novakovic R, O'Kiely D, Thomson S, Mitchell S, Herterich R (2016) Bananas—defects in the jet stripping process. In: Proceedings of European study group with industry. Dublin, Ireland

Hocking GC, Sweatman WL, Fitt AD, Roberts M (2010) Coating deformation in the jet stripping process. In: Marchant T, Edwards M, Mercer G (eds) Proceedings of 2009 mathematics-in-industry study group. Wollongong, Aust., ISBN: 978-1-74128-181-1

Hocking GC, Sweatman W, Fitt AD, Breward C (2011) Deformations during jet-stripping in the galvanizing process. J Eng Math Tuck Special Issue 70:297–306. https://doi.org/10.1007/s10665-010-9394-8

Howison SD, King JR (1989) Explicit Solutions to Six Free-Boundary Problems in Fluid Flow and Diffusion. IMA J Appl Math 42:155–175. https://doi.org/10.1093/imamat/42.2.155

Thornton JA, Graff HF (1976) An analytical description of the jet-finishing process for hot-dip metallic coatings on strip. Metall Mater Trans B 7B:607–618. https://doi.org/10.1007/BF02698594

Tuck EO, Bentwich M, Van der Hoek J (1983) The free boundary problem for gravity-driven unidirectional viscous flows. IMA J Appl Math 30:191–208. https://doi.org/10.1093/imamat/30.2.191

Tuck EO (1983) Continuous coating with gravity and jet stripping. Phys Fluids 26(9):2352–2358. https://doi.org/10.1063/1.864438

Van Dyke M (1975) Perturbations methods in fluid mechanics. Parabolic Press, Stanford, California, p 0-915760-01-0

# Measuring Moisture in Bauxite with Microwaves

**Mark McGuinness, Sean Bohun, Vincent Cregan, William T. Lee, Jeff Dewynne, Gary O'Keeffe, and Tuoi Vo**

## 1 Introduction

Rusal Aughinish Alumina Ltd. (RAAL) located on the Shannon Estuary (see Fig. 1) refine bauxite into aluminium oxide. Also called alumina, it is a critical component in the production of aluminium. The refinery is the largest in Europe and one of the world's most energy efficient. The facility includes a deepwater terminal on the Shannon estuary that facilitates efficient deliveries to the plant and shipping of alumina to smelters across Europe. The Bayer process is used to extract the alumina from a reddish brown ore called bauxite. The main components of dry bauxite are alumina (30–50%), iron oxide, silicon oxide, and titanium oxide.

### *1.1 Problem Description*

One of the key concerns of RAAL is the precise moisture content of the bauxite before it is processed at the plant. The moisture content is the amount of water naturally absorbed by the bauxite, the measurement of which is necessary to determine the mass of the bauxite alone, referred to as dry mass. Bauxite suppliers blend dry and

M. McGuinness (✉)
Victoria University of Wellington, Wellington, New Zealand
e-mail: Mark.McGuinness@vuw.ac.nz

S. Bohun
University of Ontario, Oshawa, ON, Canada

M. McGuinness · V. Cregan · G. O'Keeffe · T. Vo
University of Limerick, Limerick, Ireland

W. T. Lee
University of Huddersfield, Huddersfield, UK
URL: https://pure.hud.ac.uk/en/persons/william-lee

J. Dewynne
University of Oxford, Oxford, UK



R. McKibbin et al. (eds.), *Proceedings of the Forum "Math-for-Industry" 2019*,
Mathematics for Industry 36, https://doi.org/10.1007/978-981-19-1154-5_2

**Fig. 1** Aerial view of Rusal Aughinish Alumina Ltd. plant

wet bauxite to meet target specifications. With each shipment of bauxite, the supplier issues a certificate stating the moisture content of the shipped material. The moisture content of bauxite is very important especially when it comes to cost. For example, if the moisture content by weight is 5%, then one expects to pay 95% of the total weight for the bauxite. However if laboratory tests show that there is in fact 10% moisture, then one would be paying an extra 5% for the bauxite, and thus losing money. RAAL would like to carry out their own independent survey of the moisture content of the entire cargo. Thus, RAAL need to know the moisture content of the shipped bauxite.

Samples taken from the conveyor belt are taken to a laboratory, and the loss of weight on drying technique is used to directly measure moisture content. RAAL collect three samples of bauxite per day from the conveyor belt (see Fig. 2) that carries the bauxite from the ships to the initial grinding stage. These samples are then taken to the laboratory where they are weighed, dried and weighed again. The moisture content $M$ is calculated by comparing their initial wet mass to their final dry mass. This is called the weight percentage on a wet basis moisture content. It is given by

$$M = \frac{W_W - W_D}{W_W} \times \frac{100\%}{1}, \tag{1}$$

where $W_W$ is the weight of the initial wet sample and $W_D$ is the weight of the sample after drying. Whilst, this method gives accurate measurements, the method is tedious and time-consuming as the bauxite has to be sampled, handled and dried. Furthermore, it gives no information on what the moisture content was at times between the measurement times.

To improve the frequency of the moisture content measuring process, RAAL installed a moisture analyser, which saves readings once every second. It looks very

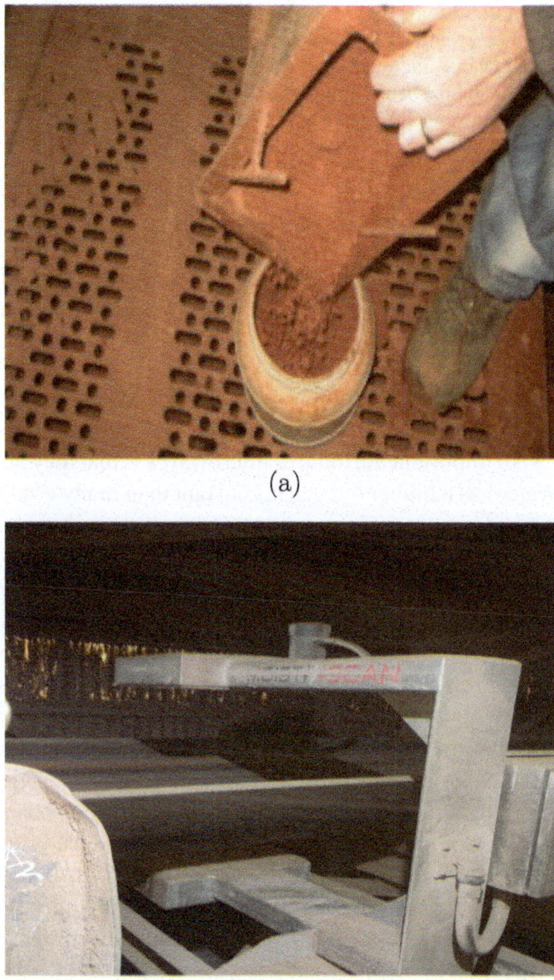

(a)

(b)

**Fig. 2** **a** Sampling bauxite from belt and **b** water moisture analyser

similar to the device reported and sketched in Viana (2013), who discusses the use of two different microwave frequencies, and the efficacy of linear regressions for obtaining moisture content.

The device uses microwaves to ascertain the moisture content in the bauxite. Microwaves are electromagnetic waves with frequencies in the range 0.3–300 GHz. When a microwave passes through the bauxite, its energy is partially absorbed by the bauxite, causing:

- *Attenuation*—a decrease in wave intensity;
- *Phase shift*—an offset in wave displacement.

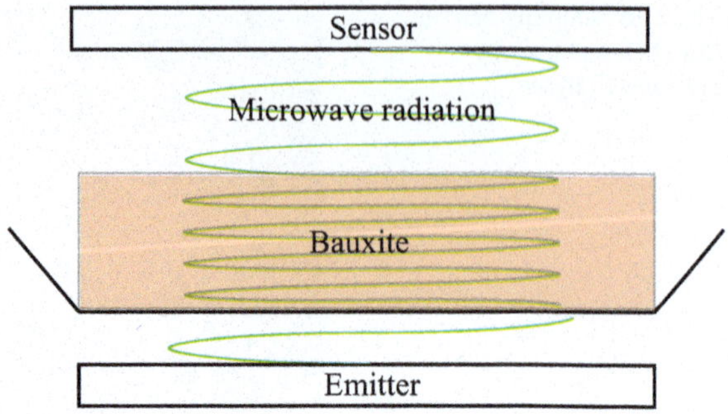


**Fig. 3** Schematic of analyser. Microwaves pass from lower emitter, through the belt and the bauxite and upward towards the upper sensor

The degree to which the wave is attenuated and phase-shifted by the bauxite depends on its electrical properties (e.g. dielectric constant) and the wave frequency.

An important attribute of microwaves is that they are highly interactive with water. Water has a higher dielectric constant than many common materials giving hope that the transmission of microwaves through moist bauxite might provide a good signal-to-noise ratio for measuring moisture content.

A schematic of the analyser at RAAL is shown in Fig. 3. Once every second, a low power microwave signal passes from the lower, emitting antenna, passes through the bauxite, is attenuated and phase-shifted and is then received by the top antenna. By comparing the received signal with the emitted signal, the analyser can determine the overall attenuation and phase shift of the microwaves due to their interactions with both the material and its water content. The device records several measurements, the most important being phase angle 1, phase angle 2 and the signal attenuation. We note that phase 1 is the difference between phases of a reference signal and the received signal. Phase 2 is similar, but the received signal is shifted by approximately 90°. A separate channel is used for phase 2. The bauxite depth (or level) and bauxite tonnage per hour are also recorded. Based on these measurements an algorithm in the analyser is used to calculate the moisture content once every second.

## *1.2 Problem Statement*

In theory, the online analyser should allow RAAL to produce a profile of the moisture content of a ship in real time and thus provide an early indicator for high moisture cargos, as well as almost continuous measurements of moisture content. However, at present there are discrepancies between the moisture content measured directly in samples taken to the on-site laboratory and the measurements given by the analyser.

The key questions for the Study Group were:

- Is the analyser capable of measuring moisture accurately (within $+/-$ 1% moisture on a wet basis)?
- What is the best way to use the data that is output by the analyser?

- Is the measurement from the analyser bauxite-dependent? RAAL currently process four different known types of bauxite. Will the analyser need to be reconfigured for a particular type of bauxite?
- Whilst it is believed that the phase shift is the most significant parameter, can the attenuation be used to fine-tune the moisture content measurement of the analyser?
- Is a density function required?

RAAL have three months' worth of data measured at one second intervals from the analyser, whilst in the same period laboratory data is time stamped with approximately three samples per day. These two sets of data will be used throughout the report to see if it is possible to construct a calibration model that would take the readings from the microwave analyser, and correctly predict the moisture content of the bauxite.

We begin by reviewing previous studies relevant to this challenge. Then we investigate the raw phase data provided during the microwave measurement process. Based on the information gained from our analysis of the raw data, we reduce the associated noise, present a new algorithm for inferring phase shift and briefly investigate the effectiveness of phase shift and attenuation in predicting moisture content. Then we review dielectric polarisation theory, specialised to the question of moisture content in bauxite with varying bauxite depths. We explore how this theory relates to the analyser problem at RAAL. A simple Beer Law model for the relationship between attenuation and moisture content is explored in Sect. 6.

## 2 Previous Studies

There is very little published on analysing microwave propagation through bauxite. The Features paper by Viana (Sidney (2013) describes in broad terms the use of a microwave analyser for providing frequent measurements of moisture content in bauxite travelling on a conveyor belt. This analyser appears to operate in a similar manner to the analyser that provided the data in the present paper. They describe a linear regression used to infer moisture content assuming a linear dependence on

1. attenuation divided by some measure of wet mass, and
2. phase shift divided by some measure of wet mass.

A reasonable proxy for wet mass would be either the height (that is, depth) of bauxite, or the height times a measured density inferred from a weight measurement made on the bauxite as it travels along the conveyor.

Viana describes an extended regression model, still linear, when two different microwave frequencies are used. They mention that calibrating the regression models is complicated and time consuming, which is a problem if frequent re-calibration is required. This echoes the concerns expressed by RAAL at our Study Group. Their laboratory values have a sample moisture variance of about 0.7%, and they obtained an analyser measurement variance of 0.9% after rejecting some outliers. Hence their analyser results are accurate to within ±2% moisture content wet basis, using a 95% confidence interval, which is about half the accuracy RAAL are hoping for. Viana emphasises the need for care when taking and transporting samples for lab measurement of moisture content, crucial for the calibration of the regression model.

There have been more studies of the microwave permittivity of grains and seeds, used to determine moisture content and density, many associated with (Nelson and Trabelsi 2010; Trabelsi et al. 2001, 2000, 1998; Nelson 1987; Kraszewski and Kulinski 1976). Referring in particular to Nelson and Trabelsi (2010), a complex permittivity is the usual starting point and the focus of study, $\epsilon = \epsilon' + i\epsilon''$, where $\epsilon'$ is the usual real permittivity, and $\epsilon'' = \sigma/\omega$, where $\sigma$ is the electrical conductivity and $\omega$ is the angular frequency of the microwave signal. They note that there is dissatisfaction with measurement reliability in grain and seed, especially at higher moisture levels. They introduce a moisture calibration function

$$\psi = \sqrt{\frac{\epsilon''}{\epsilon'(a_f\epsilon' - \epsilon'')}}, \tag{2}$$

where $a_f$ is a calibration constant, obtained as the slope of a graph of $\epsilon''/\rho$ versus $\epsilon'/\rho$, where $\rho$ is a density or some other measure of how much material the microwaves are passing through. The values of $\psi$ are to be used as the linear regression variable that determines moisture content. Largely empirical arguments for the form of $\psi$ are presented in Trabelsi et al. (1997), to motivate dividing the ratio of the imaginary and real components of permittivity by a measure of density, and taking the square root of $\psi$ is justified by the further straightening effect it has upon data gathered for a variety of grains and seeds. They also note that the ratio of attenuation to phase shift should be a good measure of moisture content, since it is independent of density (or in our case bauxite height), but they present data showing that this ratio is significantly nonlinear when plotted against moisture content.

An issue common to both of the above research areas is nonlinearity of the relationships between phase shift, attenuation, or combinations and ratios of these, when plotted against measured moisture content. This nonlinearity gives issues at higher moisture content, if the analyser has been calibrated over a range of lower moisture values.

## 3 Analysis of Raw Phase Data

In this section we analyse the raw phase data from microwave measurements, and we leverage the resulting understanding to reduce noise in that data. A description of how the phase shifts are measured by the analyser is given in Sect. 3.1. In Sect. 3.2 we discuss how noise rounds the corners of the squircle (i.e. plot of phases $\phi_1$ and $\phi_2$ against each other) obtained from the data. In Sect. 3.3 we discuss how to improve the raw phase data by projecting the data onto a theoretical parallelogram. This may be compared with the simpler approach of accepting the squircle data without projection, as outlined in Appendix B. Section 3.4 explains the relationship between the sudden jumps in the phase values and the changes in bauxite depth. In Sects. 3.5–3.8 we outline a phase lifting method to obtain the correct phase values. The resulting lifted values are then compared to the raw data. In Sect. 3.9 we describe the effect that the small time lag during rapid changes in height has on the lifted phase versus depth plot. The relationship between attenuation and depth is approximated via a sinusoidal type-expression in Sect. 3.10. In Sect. 3.11 we use the analysis from the preceding sections to introduce an expression for the moisture content.

### *3.1 Phase Shifts*

The measurement of phase shift appears to be crucial to the success of the microwave instrument being tested by RAAL. The microwave analyser is intended to measure the moisture content of bauxite every second, as it is being offloaded from a ship. Supplemented by lab measurements for calibration confirmation purposes, these measurements are intended to provide an essentially continuous measurement of moisture content.

The details of how the phase is computed are confidential to the company that makes the moisture analyser. However, it is possible to make some conclusions based on conversations with a vendor representative, and on the data that comes out of the analyser.

Two raw phase values are provided by the analyser, which we will call $\phi_1$ and $\phi_2$. The first phase is obtained by comparing a reference signal (which we will think of as the transmitted microwave signal, fed to a processor by an internal pathway, that is, by wires) with the signal received after the transmitted microwaves have passed directly through the bauxite that is travelling on a conveyor belt between transmitter and receiver. How this comparison is performed may be deduced from the literature and from the data that comes from the analyser. We know that both signals have the same frequency, 900 MHz. We also know that $\phi_2$ is intended to be lagged by approximately $\pi/2$ from $\phi_1$, by passing the microwave signal from the receiving antenna through a circuit that is separate from that producing $\phi_1$ and is designed to produce this phase difference. The reason for this will become clear. The actual phase difference indicated by plotting $\phi_2$ against $\phi_1$ is about 68°. This is clear when a plot of $\phi_2$ against $\phi_1$ is compared with plots of $\arcsin(\sin(\phi))$ against $\arcsin(\sin(\phi + \Delta\phi))$, for various values of the phase difference $\Delta\phi = \phi_1 - \phi_2$.

After studying the analyser data and researching the various common ways to calculate the phase difference, $\Delta\phi$, between two signals, we conclude that the two signals are being clipped, passed through an XOR gate, and the output averaged. Clipping turns a sinusoid into something close to a square wave, preserving the times of zero crossings; that is, the times at which the sinusoids pass through zero in value. The wave has two states, like on and off, or true and false. XOR then returns

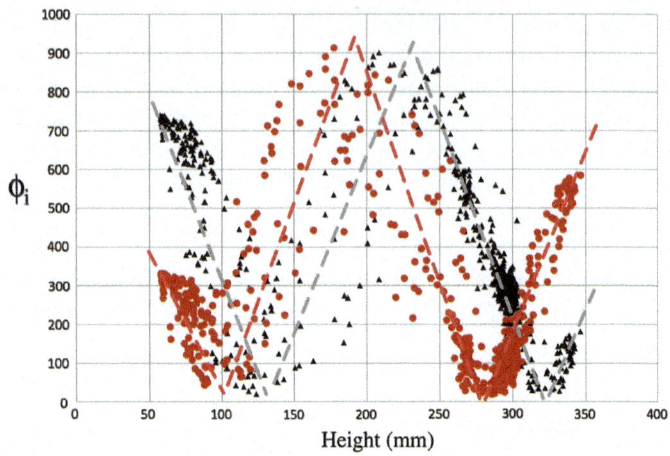


**Fig. 4** Plots of $\phi_1$ (red disks) and $\phi_2$ (black triangles) versus height of bauxite, during a time period of about 30 min, from the ship Maia. The dashed lines are a rough guide to the theoretical values of $\phi$. There is a puzzling extra feature here, also visible in the squircle plots (see also Fig. (11) for a clearer indication of this from a longer time series), the appearance of two different trend lines for phase versus depth, like in hysteresis. That is, especially in the depth range 100–300 mm for the red disks, there seem to be two lines, one above and one below the dashed fit

zero during times when both waves are in the same state, and one during times when one wave is on and the other is off. In the absence of noise, then, if both waves are perfectly in phase (a phase difference of zero), the output from the XOR gate is always zero. If the two waves are $\pi$ out of phase, the XOR output is always one. For phase differences between zero and $\pi$, the average of the XOR output is proportional to the phase difference. When this difference begins to exceed $\pi$, we can see that the average XOR output declines, reaching zero when the phase difference reaches $2\pi$. In this way, in the absence of noise, a sawtooth wave is produced that has a period $2\pi$.

The amplitude of the sawtooth that is obtained by the above-described XOR process lies naturally in the range [0, 1] and is dimensionless. For noise-free sinusoids, the phase shift that corresponds to the value one would be $\pi$ radians, also dimensionless. This signal is scaled by the equipment to a dimensionless range that is close to [0, 1000], and the value 1000 might then be regarded as meaning a phase shift of $\pi$ radians.

The second phase difference $\phi_2$ is produced in the same way, after shifting the received signal by a phase of about $\pi/2$, to produce a second sawtooth wave. Views of these waves are plotted from analyser data in Fig. 4, using the depth of bauxite as a proxy for moisture content, and show the sawtooth nature of the signals. The deeper the bauxite, the more moisture is in the path of the microwaves, if we ignore changes in moisture fraction.

Figure 5 shows the result of plotting the values of $\phi_1$ and $\phi_2$ in Fig. 4 against each other. Instead of the expected parallelogram or square, we see a parallelogram or

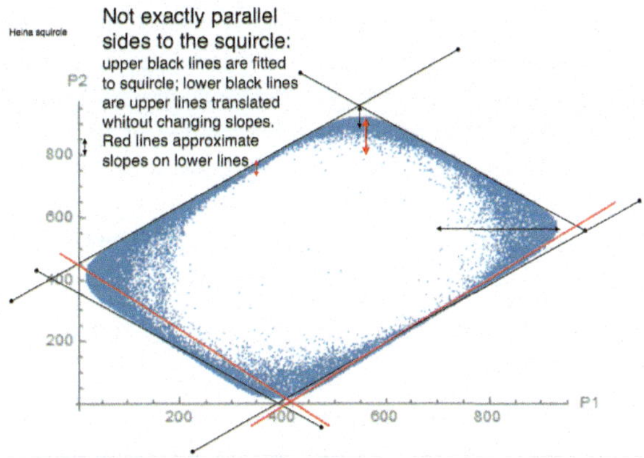


**Fig. 5** Plots of $\phi_1$ versus $\phi_2$ during the unloading of the ship Heina. The upper two black lines were drawn parallel to the upper straight parts of the squircle. The lower two black lines are translated versions of the upper two, showing that the slopes change a little from top to bottom and left to right. The red lines approximate the true slopes of the lower two squircle edges. The red double-headed arrows indicate variance times four, at different parts of the squircle. The black double-headed arrow indicates the apparent shift of the measured mean phase from true mean. The time period was over several days

square with rounded corners, termed a *squircle*. Noise is responsible for the rounding of the corners, as will be explored in Sect. 3.2. Apparently there is more variance in a phase when it has mean values near zero and $\pi$, than elsewhere.

## 3.2 *Noise and the Squircle*

We now understand that the cause of the rounding of corners is due to noise. This is clear from two arguments. One argument is to consider the effect of noise on an XOR signal that comes from two clipped signals (the reference signal and one of the received signals that has passed through the bauxite) that are in phase, which in the absence of noise has a zero average value. Noise will jitter the clipped inputs, giving a non-zero average from the XOR gate due to the inputs never quite matching. That is, noise will bias the average measured phase away from zero, causing these values to veer away from the theoretical straight line near zero. The same thing happens near the location where the two signals being compared have a phase difference of $\pi$—the XOR signal will be biassed away from the average value of one by the jitter in the input signals. This effect will be seen in both $\phi_1$ and $\phi_2$, since they are both measured in this manner.

A second argument is to consider probability distributions on finite domains, since the clipping and XOR process is physically constrained to return a fixed range. It is sufficient for our purposes to consider distributions on semi-infinite domains. There are several continuous ones, one being the folded normal distribution, and another being the gamma distribution. When the true mean (that is, the true phase in our setup) is zero, the folded normal distribution becomes the half-normal, with a mean value that is $\sqrt{2/\pi}$ times $\sigma$, where $\sigma^2$ is the variance of the normal distribution. When the XOR signal is averaged, the resulting phase is the mean of the phase and is shifted by an amount that is approximately equal to $\sigma$. A gamma distribution has a similar property that the mean is shifted by an amount proportional to $\sigma$ (the constant of proportionality is $\beta$, a parameter in the gamma distribution).

### 3.3 Corrections to Phase Values

We now seek to leverage the knowledge discussed above, to find a good way to improve the raw phase data provided by the analyser. It is clear that since the mean phase is biassed near corners, it is desirable to smooth and correct the average phase values by projecting from the raw squircle data onto the parallelogram that follows the theoretical mean phase values in $\phi_1$, $\phi_2$ space that are not noise biassed. This is the shape ABCD sketched in Fig. 6.

The suggested process begins with a fit of the parallelogram ABCD to the phase data. This fit is here done by identifying by eye the eight points a, b, c, d, e, f, g, h, along the parallelogram that correspond to straight sections. These points are indicated in Fig. 7. Since we anticipate that squircles depend only on the electronic setup and lags, they will likely have the same shape from ship to ship and for different bauxites, provided that the analyser setup is not altered, it is probably good enough to use the general values in Table 1.

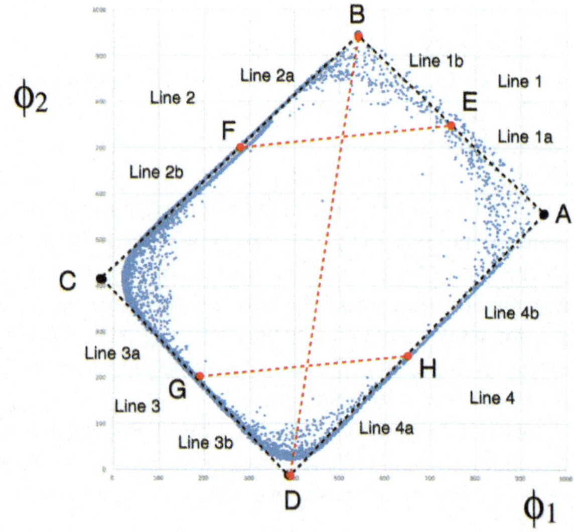


**Fig. 6** The squircle, the parallelogram ABCD of theoretical mean values of $\phi$, and labelled points used to decide how to project Maia data onto the parallelogram. The sublabelling of Lines 1–4 reflects the notation used in the MATLAB code that produces Figs. 9 and 8

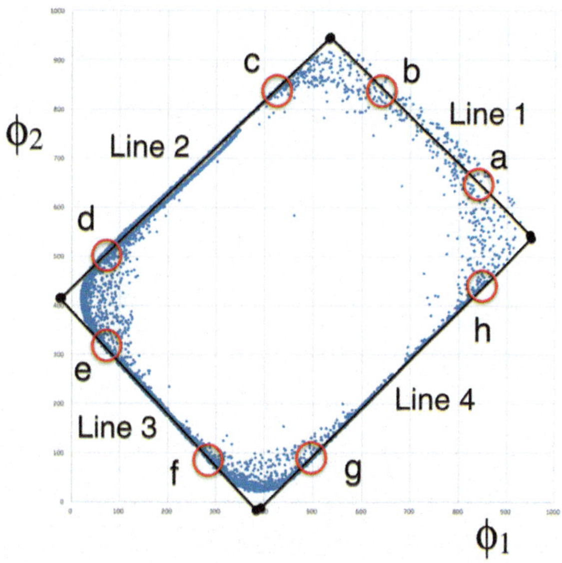


**Fig. 7** The squircle, the parallelogram of theoretical mean values of $\phi$, and labelled points used to choose straight portions of the squircle to fit straight lines to, for Maia data

**Table 1** Table of suggested $\phi$ values used to restrict the data range and fit straight lines to calculate the parallelogram behind the squircle

| Point | $\phi_1$ | $\phi_2$ |
|---|---|---|
| a | 850 | 650 |
| b | 650 | 850 |
| c | 425 | 850 |
| d | 75 | 500 |
| e | 75 | 300 |
| f | 300 | 75 |
| g | 500 | 75 |
| h | 850 | 450 |

The line through the points A, B (Line 1) is calculated by regressing $\phi_2$ against $\phi_1$, for data values that lie in the range (a, b). The values (a, b) are used only to select the data range to be fitted to.

We seek to avoid the curved parts of the squircle in this fitting, since they are biassed away from the true phase values by noise. The regression should be one that is robust to the existence of errors (noise) in both variables $\phi_2$ and $\phi_1$, for example, an orthogonal regression, also known as Deming regression, that minimises the sum of squared distances from data points to the fitted line, rather than the usual minimisation of squared differences in $\phi_2$ values between points and line. We use the singular value decomposition to achieve orthogonal regression in the MATLAB code associated with this work.

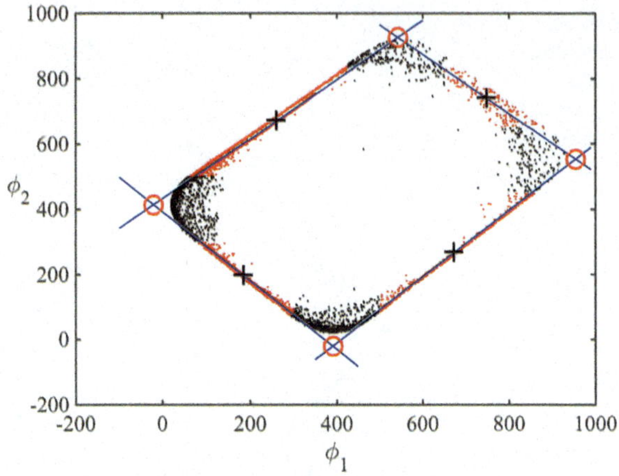


**Fig. 8** The squared squircle, for Maia phase data, showing the parallelogram of theoretical mean values of $\phi$, obtained using MATLAB to fit straight lines to red-coloured points, chosen as sections that are relatively straight. Fits use singular value decomposition to do an orthogonal regression. Vertices computed are identified using circle symbols. Centres of the sides of the squared squircle have also been calculated and are shown as plus signs

The other three lines are obtained similarly, to obtain Line 2 through points B, C; Line 3 through points C, D; and Line 4 through points D, E. Next their intersections are calculated as the points A, B, C, D. For example, if Line i is given as

$$\phi_2^i = m_i \phi_1^i + c_i \ , \quad i = 1..4 \ ,$$

the intersection point B between Lines 1 and 2 has coordinates $B = (\phi_1^B, \phi_2^B)$, where

$$\phi_1^B = \frac{c_2 - c_1}{m_1 - m_2} \ ,$$

and

$$\phi_2^B = m_1 \phi_1^B + c_1 \ .$$

The result of coding the above approach to Maia phase data is shown in Fig. 8. Vertices C and D have a component of order $-20$.

In the event that some values of A, B, C, or D lie outside the range (0,1000), a rescaling of all phase values can be conducted, although there is no harm in ending up with a different range after projection, with possibly some negative phase values resulting. If this is of concern, it is easy to rescale all projected phi values, for example to $[0, \pi]$, by taking

$$\phi_{\text{new}} = \pi \left( \frac{\phi - \phi_{\text{min}}}{\phi_{\text{max}} - \phi_{\text{min}}} \right) ,$$

where $\phi_{\text{min}}$ is the minimum of the original $\phi$ values, and $\phi_{\text{max}}$ is the maximum.

Next we compute the midpoints on each Line, as labelled E, F, G, H, in Fig. 6. For example, the vector E = (A + B)/2, so just take the average of each component of A and B to get the components of E. A simple way to correct onto the parallelogram ABCD, implemented in MATLAB, is as follows. There are four cases to consider:

1. For all points that lie above the line through EF, we will correct $\phi_2$ values (these will be the ones near $\pi$, i.e. 1000) by projecting along constant $\phi_1$ values (since $\phi_1$ has a relatively small variance there and is not biassed in its mean value by being up against an end-value). We project onto one of Line 1 or Line 2. The choice of which Line to project onto is: if $\phi_1 < \phi_1^B$, where $\phi_1^B$ is the $\phi_1$ value at point B, use Line 2; otherwise, use Line 1.
   This projection is obtained by keeping the $\phi_1$ value of the data point, but change the $\phi_2$ value so that it satisfies the equation for Line 2 (or Line 1, if the data point lies to the right of B). This will smooth the data, using the most reliable $\phi$ value. Smoothed phase points will now lie on one of the lines AB or BC. There will still be error or noise in the data, but it should be much reduced.
2. For all points that lie below the line through GH, correct $\phi_2$ values by projecting along constant $\phi_1$ values, onto one of Line 3 or Line 4. If $\phi_1 < \phi_1^D$, choose Line 3 to project onto; otherwise, choose Line 4.
3. For all unprojected points that lie to the left of the line through FG, the idea is to correct $\phi_1$ values by projecting along constant $\phi_2$ values. In practice, in order to project all points in the squircle, we work with all points to the left of the line BD. If $\phi_2 < \phi_2^C$, and provided $\phi_2$ lies above the line GH, choose Line 3 to project onto; otherwise, If $\phi_2 > \phi_2^C$, and provided $\phi_2$ lies below the line EF, choose Line 2 to project onto. The formulae for these lines will need to be rearranged to make $\phi_1$ the result, given the fixed value for $\phi_2$.
4. For all unprojected points that lie to the right of the line through BD, correct $\phi_1$ values by projecting along constant $\phi_2$ values. If $\phi_2 < \phi_2^A$, choose Line 4 to project onto but only for points above the line GH; otherwise, choose Line 1, for points below the line EF. The formulae you have for these lines will need to be rearranged to make $\phi_1$ the result, given the fixed value for $\phi_2$.

This outline of a simple way to correct onto the parallelogram ABCD is the basis of the MATLAB code. The projections are straightforward if a little tedious. Figure 9 shows the raw and projected $\phi$ values after the process outlined above is halfway through. Figure 10 shows the original raw phase data in black forming a squircle and the result of squaring the squircle as outlined above as a red line.

The result of projecting the raw phases onto the parallelogram and then rescaling the resulting phase values onto $(0, \pi)$ may be observed versus bauxite height in Fig. 11. This does not look very different from a plot of raw phase values versus height. The projected phases have slightly sharper vertices, but noise disguises this. In both raw and projected phase data, there is a puzzling hysteresis effect visible at intermediate depths.

As moisture content varies by a much smaller amount than height, the height becomes a reasonable proxy for the true phase angle. So a plot of $\phi_1$ against height gives us a good idea of how the true phase should vary with the total moisture (or

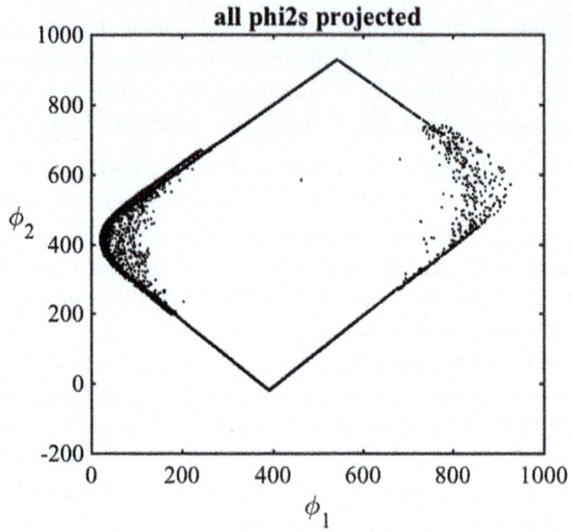


**Fig. 9** An illustration of the process of squaring the squircle, for Maia phase data, showing the effect on raw phase data after the process is half finished. Points above the line through E, F and points below the line through G, H have been projected vertically, altering their $\phi_2$ values but preserving their $\phi_1$ values

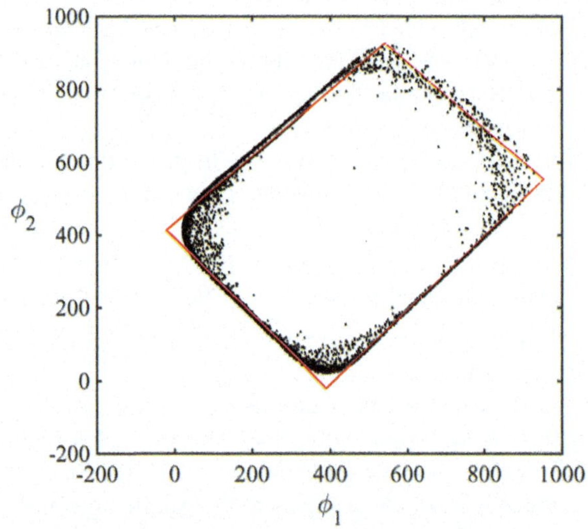


**Fig. 10** The original raw phase data are shown before projection in black, and the projected phase data are shown in red, in an illustration of the process of squaring the squircle, for Maia phase data. Points above the line through E, F and points below the line through G, H have been projected vertically, altering their $\phi_2$ values but preserving their $\phi_1$ values. Points between the line through E, F and the line through G, H have been projected horizontally, to the right if they are to the right of the line segment BD, otherwise to the left. Note that the projected data, plotted as single points, do indeed fall upon the straight projection lines illustrated in Fig. 6

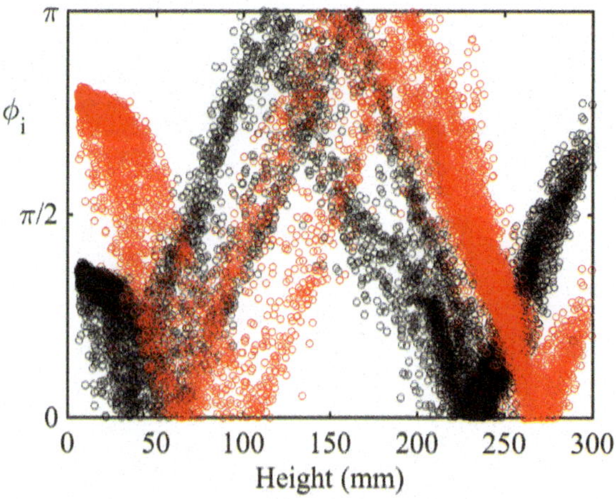


**Fig. 11** A plot of projected phases versus net bauxite height, for data from the ship Maia, using all of the data (over 27,000 points). $\phi_1$ in black, $\phi_2$ in red

number of polarisers due to water molecules) in the scanned region. We see a total change in phase of more than $3\pi$ from zero moisture to maximum moisture. We also see a lot of "noise", departures from the clean zig-zag that would arise if moisture content was constant.

## 3.4 Data Exploration

We noticed what we at first thought were transients in the phase values $\phi_1$ and $\phi_2$, which are clearly visible in Figs. 12 and 13 whenever the bauxite depth changes from zero (which is about 50 on the scale shown) to non-zero values. These jumps in phase value last about 6 or 7 s, which is the time scale over which the depth is changing. The jumps at first sight appear to be as much as $\pi$.

However, inspection of Fig. 4 (or Fig. 11) reveals what is occurring—that figure makes it clear that phase has changed by more than $2\pi$ during the transition time from zero height to a height more than 300 mm in just four seconds. A close examination of the changes in $\phi_1$ and $\phi_2$ versus height as presented in Fig. 14 only just manages to reveal what is actually happening—phase values are quickly moving through more than $2\pi$, on the same time scale as the bauxite depth is changing. Since there are only four values, the resulting phase changes are only just resolved. This allows an algorithm of moving around the squared squircle to be used to compute total phase change from the zero bauxite case, provided that bauxite depth does not change any more rapidly than it has in this example, increasing from zero to 240 mm over a period of about four seconds. Since the belt moves at about 1.88 m/s, that is over a distance of about 7 m. A more gradual increase in bauxite heights, that extends over 15–20 ss instead of four, would make the computation of phase angle more reliable.

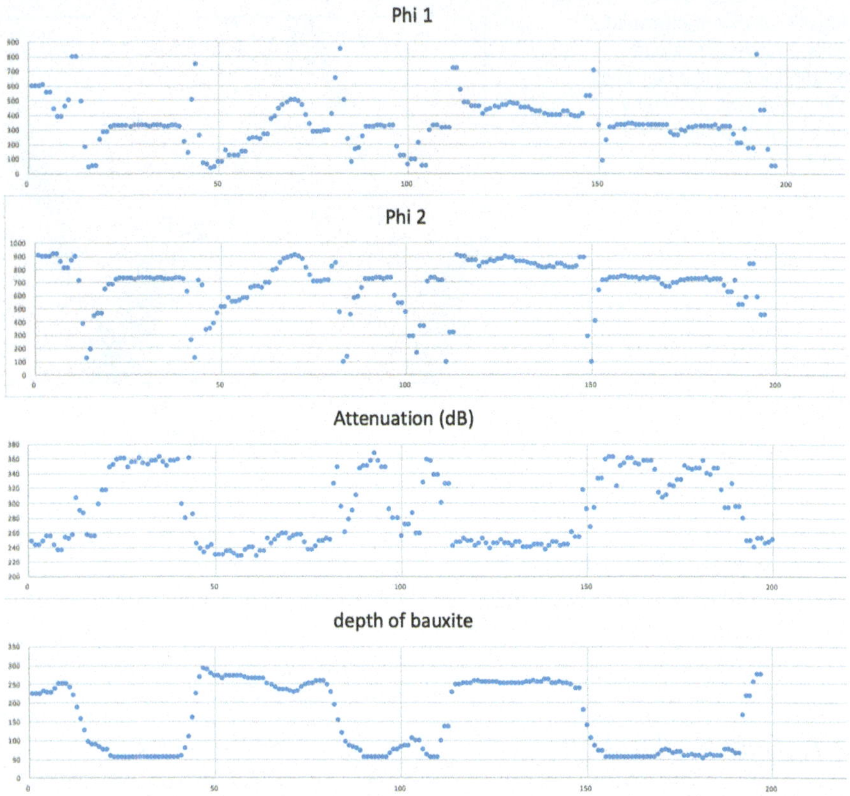


**Fig. 12** Plots of raw $\phi_1$, raw $\phi_2$, attenuation, and bauxite height versus time in seconds, for data from the ship Heina. Heights are not zeroed here. Values of $\phi$ are as provided by the equipment, where the value 1000 corresponds to $\pi$ radians, both being dimensionless

If bauxite height increases (or decreases) any faster than 240 mm in four seconds, then zero height measurements are not a lot of help for calibration, and moisture measurement must rely instead on calibration to lab samples taken at the new height. Any changes in height of more than 200 mm in less than four seconds will jump the phase through more than $2\pi$ too rapidly to be visible to any algorithm for determining phase based solely on the two measured phase angles.

## 3.5 *Lifting the Phase*

A single phase value measures angles between zero and $\pi$, and cannot measure phase values beyond $\pi$ without further information. Having two phase data values with a lag between them, as noted in Sect. 3.1, allows calculation of the total change in phase up to $2\pi$, which is once around the squircle. Between one measurement of $\phi_1$

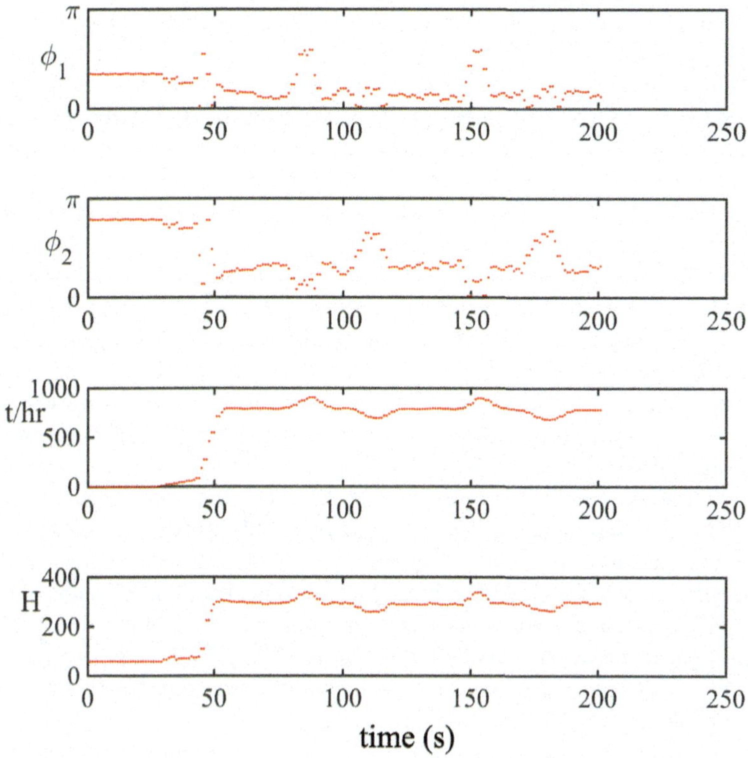


**Fig. 13** Plots of $\phi_1$, $\phi_2$, tonnage, and bauxite height versus time in seconds, for data from the ship Maia. Heights are not zeroed. Note that the tonnage has been lagged by four seconds to match changes in H for Maia data

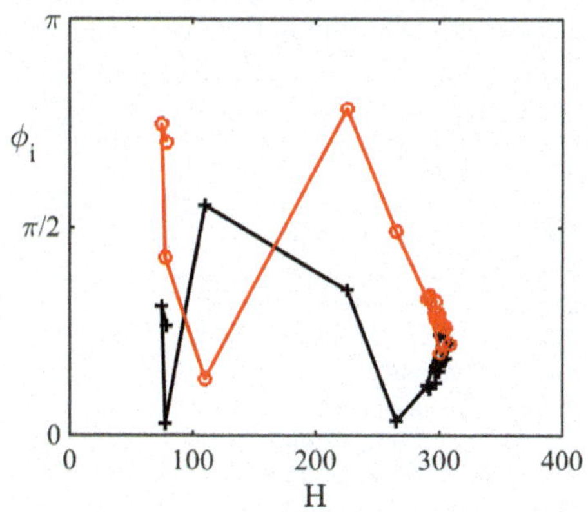


**Fig. 14** Expanded plots of $\phi_1$ and $\phi_2$ versus bauxite height during the first time in the previous plot (13) that the bauxite height first jumps up from zero

and the next, the correct value of true phase may be the apparent value plus or minus an integer multiple of $2\pi$. We see jumps in phase values as judged by the bauxite depth value of more than $\pi$, for example, in Fig. 14.

So even with two phase values, the phase angle produced is ambiguous, or multivalued, just like the phase of complex numbers plotted in an Argand diagram. One challenge faced by this study group is to produce a unique and correct value of the actual phase shift, beyond the $2\pi$ limitation. We call this *lifting the phase*, echoing language used in circle maps in mathematics.

In experiments on biomass pellets (McKeown et al. 2017) moving through a region of fixed diameter, the density of the flow is a key measurement as it directly affects the apparent phase shift. In the present case of bauxite flowing on a conveyor belt, the key variable is the height of the bauxite, which changes from zero to a maximum around 30 cm within seconds. So the emphasis on measuring flow density in the biomass case is replaced here by an emphasis on using the separately measured height (thickness) of the bauxite.

We cannot assume that the phase difference does not jump by more than $2\pi$ between one measurement and the next, since bauxite height can and does change by a large amount from one second to the next. The only way to narrow down the change in phase and account for possible jumps of $2\pi$ is to check the current value of bauxite depth, and use an averaged sawtooth plot of phase versus depth to decide the true ("lifted") phase angle. We already know that relatively rapid changes in phase are routinely observed in practice, when new material is placed on an empty conveyor, and when bauxite supply to the conveyor is stopped.

An average of the data shown in Fig. 11 has been computed on 70 height bins, resulting in the much clearer plot shown in Fig. 15. This provides a clear basis for properly assigning a true phase angle, or lifting the phase data to a true phase shift value, in conjunction with the squared squircle, as we will now describe.

The aim is to lift each sawtooth phase plot, using the squared squircle and the smoothed sawtooth of phase versus height to decide which multiple of $2\pi$ the phase is in. A sketch of a lifted phase, together with the sawtooth it is lifting, is presented in Fig. 16.

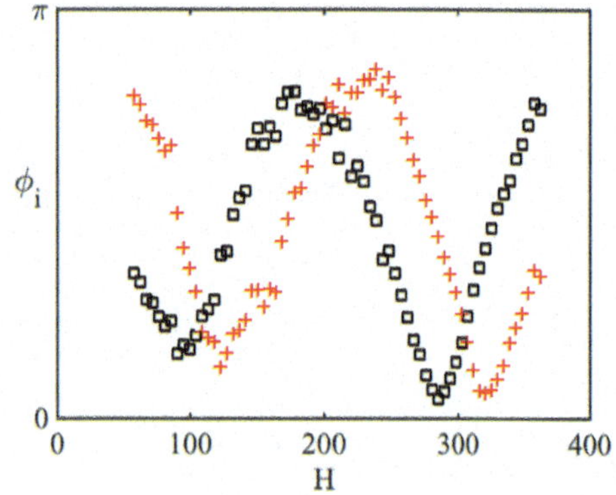


**Fig. 15** A plot of projected phases versus net bauxite height, for data from the ship Maia, using all of the data (over 27,000 points), but averaging over 70 evenly spaced bins of height values. Averaged values of $\phi_1$ are displayed using black square symbols, and of $\phi_2$ using red plus symbols

**Fig. 16** A sketch showing a noninvertible sawtooth phase (solid line), similar to a plot of $\phi_1$ versus the true phase difference from some zero value on the horizontal axis, and its lift (dashed line), that gives the true phase, on the vertical axis

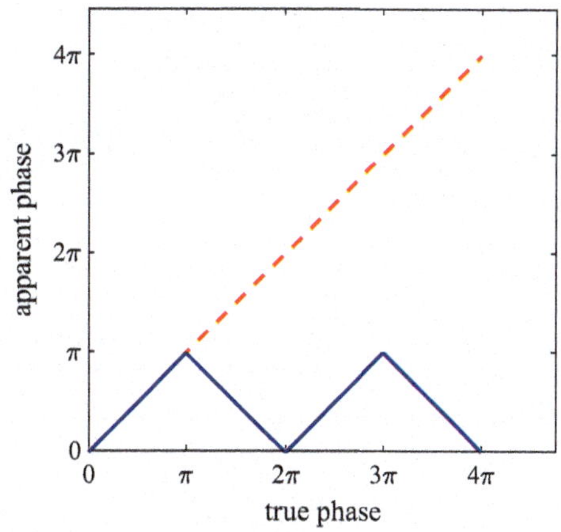


## 3.6 *Lifting Algorithm*

An algorithm for lifting to obtain the phase value is outlined here. Reference is made to the squared squircle, parallelogram ABCD, and to the annotated sawtooth graph of measured phases versus bauxite height H shown in Fig. 17.

Two values for lifted phase are obtained, $\theta_1$ based primarily on the measured values of $\phi_1$, and $\theta_2$ based primarily on the values of $\phi_2$. Bauxite height is used to help decide if phase has changed by $2\pi$ or not. Location on the squared squircle, that is, which line segment the vector $\boldsymbol{\phi} = (\phi_1, \phi_2)$ lies in, is also used to decide the lifted phase to within $2\pi$. This analysis assumes that the zero height values $\boldsymbol{\phi}^0 = (\phi_1^0, \phi_2^0)$ lie on the line segment BC. If not, the following needs to be modified.

The vertical line after location A in Fig. 17 gives the height value $H_1^{**}$, above which it is clear that lifted $\phi_1$ values are greater than $2\pi$.

The arguments that lead to formulae for the lifted phase $\boldsymbol{\theta} = (\theta_1, \theta_2)$ are typified by what follows. We begin with zero height, and we think about increasing height or increasing lifted phase, following the arrows in Fig. 18. We set the lifted phase angle to zero when there is no bauxite.

When height is small ( $H < H_1^*$, for example), and when the vector $\boldsymbol{\phi}$ is on the line segment $\boldsymbol{\phi}^\circ$ C in Fig. 18, then

$$\theta_1 = \phi_1^0 - \phi_1 \ .$$

If $H \geqslant H_1^*$, then when $\boldsymbol{\phi}$ is on the line segment $\boldsymbol{\phi}^\circ$ C

$$\theta_1 = \phi_1^0 - \phi_1 + 2\pi \ .$$

When $\boldsymbol{\phi}$ is on the line segment CD or DA, and provided height is not too big (provided $H < H_1^{**}$, for example),

$$\theta_1 = \phi_1^0 + \phi_1 ,$$

but if $H \geqslant H_1^{**}$ then

$$\theta_1 = \phi_1^0 + \phi_1 + 2\pi .$$

Note that the noise that smears the relationship between $\phi_i$ and height makes sharp inequalities based on height ineffective, but the above choices of $H_1^*$ and $H_1^{**}$ avoid this by having effect in a region that is not the parallelogram edge of current interest.

Similarly, when $\boldsymbol{\phi}$ is on the line segment AB or B$\boldsymbol{\phi}^0$,

$$\theta_1 = \phi_1^0 + \pi + \pi - \phi_1 = \phi_1^0 - \phi_1 + 2\pi .$$

We can combine these results and also allow for the lifted phase to in principle go higher than the $3\pi$ indicated by Fig. 17, as follows. We find that the lifted phase $\theta$ is given by

$$\theta_1 = \begin{cases} \phi_1^0 - \phi_1 + 2k_1\pi, & \boldsymbol{\phi} \in \text{AB} \cup \text{BC} \\ \phi_1^0 + \phi_1 + 2n_1\pi, & \boldsymbol{\phi} \in \text{CD} \cup \text{DA}, \end{cases}$$

where

$$k_1 = \left\lceil \frac{H - H_1^*}{H_{2\pi}} \right\rceil , \quad n_1 = \left\lceil \frac{H - H_1^{**}}{H_{2\pi}} \right\rceil ,$$

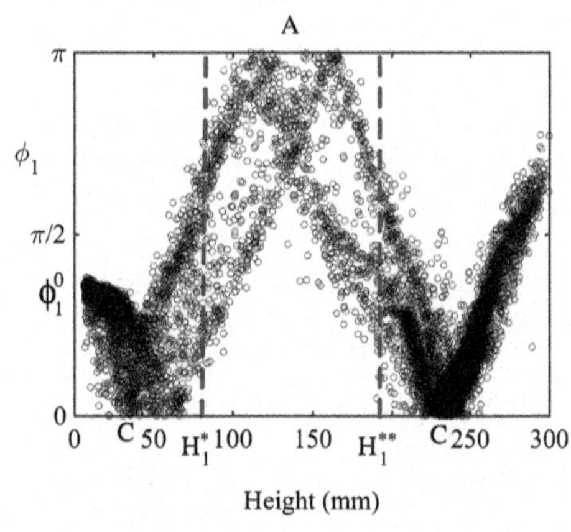


**Fig. 17** A plot of projected $\phi_1$ values vs bauxite height for the ship Maia, used to help decide if the lifted phase has gone beyond $2\pi$. The points A, C correspond to two vertices of the parallelogram in Fig. 6. The labelled values $H_1^*$ and $H_1^{**}$, which roughly bisect the rising and falling legs of the sawtooth wave, are used to help decide when lifted $\phi_1$ values have increased by more than $2\pi$

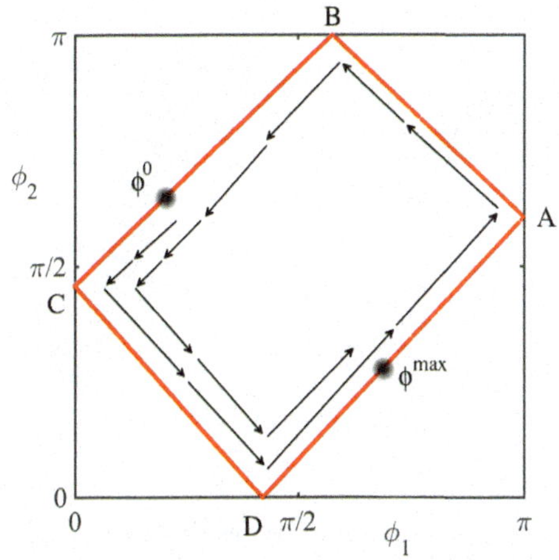


**Fig. 18** A plot of the projected $\phi$ values against each other for the Maia data, after rescaling them to the range $[0, \pi]$. The points A, B, C, D label the vertices of the parallelogram that is the squared squircle. The point $\phi^0$ labels the approximate location of the vector $\phi = (\phi_1, \phi_2)$ when bauxite height is zero, and the point $\phi^{max}$ labels the approximate location of the phase when bauxite height is maximal. The arrows indicate the direction that the vector phase travels as bauxite height increases

and $\lceil x \rceil$ is the ceiling of $x$, that is, the smallest integer that is greater than or equal to $x$. The symbol $\cup$ means the union of the two line segments it connects, so that $\boldsymbol{\phi} \in$ BC $\cup$ AB means $\boldsymbol{\phi}$ is in one of the line segments BC or AB. Note that it is clear from Fig. 19 that it is possible to choose $H_i^{**}$ halfway between $H_i^*$ and $H_i^* + H_{2\pi}$, so that

$$H_i^{**} \approx H_i^* + \frac{H_{2\pi}}{2} , \quad i = 1, 2 ,$$

where $H_{2\pi}$ is the change in height associated with a change of $2\pi$ in lifted phase, and this takes the same value for both $\phi_1$ and $\phi_2$, so it does not need a subscript or superscript for which phase it refers to.

**Fig. 19** A plot based on an extended patched version of Fig. 17, illustrating the more general situation that the sawtooth wave might go through a change in phase of more than $3\pi$. The value of $H_{2\pi}$ is illustrated, as the change in bauxite height associated with a change of $2\pi$ in the lifted phase

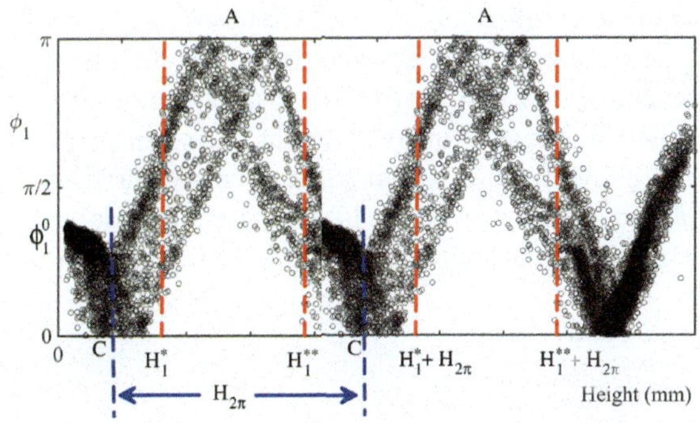

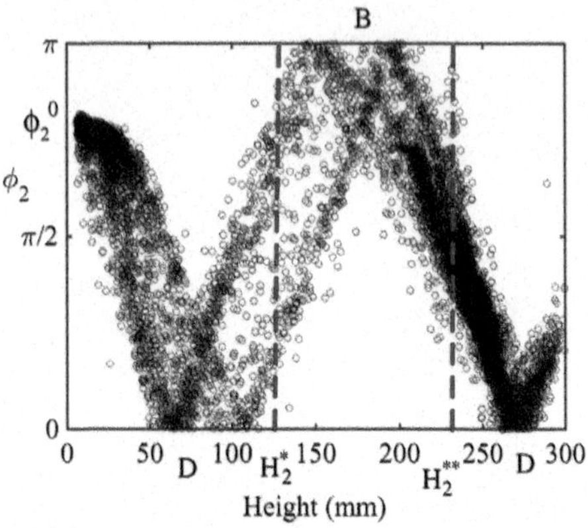


**Fig. 20** A plot of projected $\phi_2$ values versus bauxite height for the ship Maia, used to help decide if the lifted phase has gone beyond $2\pi$. The points B, D correspond to two vertices of the parallelogram in Fig. 6. The labelled values $H_2^*$ and $H_2^{**}$, which roughly bisect the rising and falling legs of the sawtooth wave, are used to help decide when lifted $\phi_2$ values have increased by more than $2\pi$

Similarly, the equation for the lifted value of $\phi_2$ is

$$\theta_2 = \begin{cases} \phi_2^0 - \phi_2 + 2k_2\pi, \ \boldsymbol{\phi} \in \text{BC} \cup \text{CD} \\ \phi_2^0 + \phi_2 + 2n_2\pi, \ \boldsymbol{\phi} \in \text{DA} \cup \text{AB}, \end{cases}$$

where (referring to Fig. 20)

$$k_2 = \left\lceil \frac{H - H_2^*}{H_{2\pi}} \right\rceil, \quad n_2 = \left\lceil \frac{H - H_2^{**}}{H_{2\pi}} \right\rceil.$$

The algorithm used to decide if $\boldsymbol{\phi}$ is in a given line segment is given in Appendix A.

The result of applying the above algorithm for lifting the phase for the Maia data is shown in Fig. 21, where the two components of $\theta$ are plotted against bauxite net height. The red dots are $\theta_2$, and $\theta_1$ is plotted using black dots. The red dots practically obliterate the black ones. Since the red dots are plotted second, this indicates that both components for $\theta$ are giving much the same phase shift values. The bilinear nature and large variance of the phases at intermediate depths are now quite clear. These correspond to very short-time periods where the depth is increasing from zero to above 200 mm, and the bilinear appearance we now understand is due to the relatively rapid changes in height during those periods. See later in this report for more investigation of this bilinear behaviour. The projected data values for $\phi_1$ and $\phi_2$ are plotted against the lifted phase values $\theta_1$ and $\theta_2$, respectively, in Fig. 22. The clean sawtooth shapes confirm the lifting algorithm is working correctly.

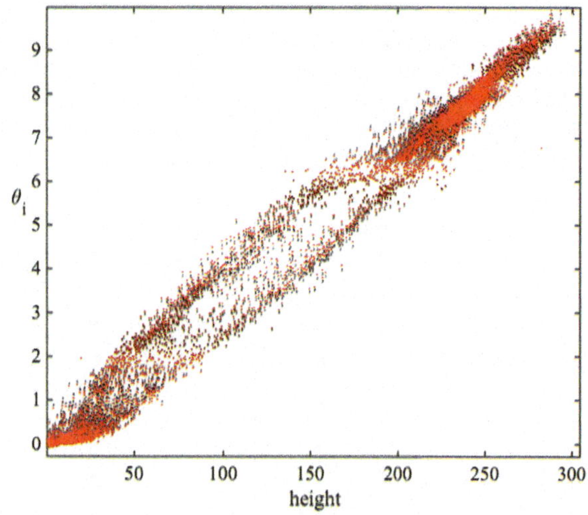


**Fig. 21** A plot of lifted phase values $\theta_1$ (black dots) and $\theta_2$ (red dots) in radians, vs bauxite height in mm, for the ship Maia

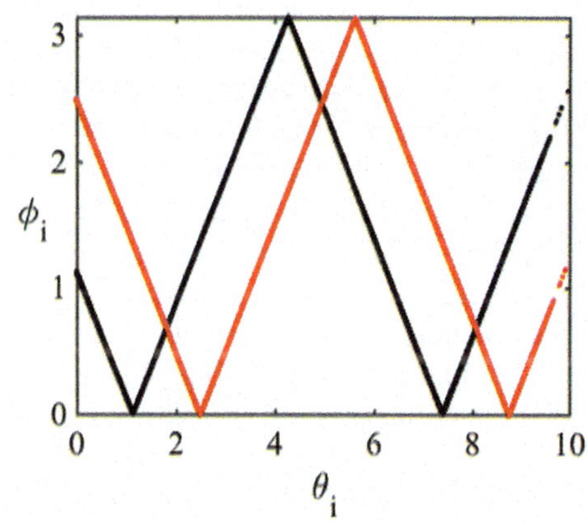


**Fig. 22** A plot of projected data values for $\phi_1$ and $\phi_2$ versus the lifted phase values $\theta_1$ (black dots) and $\theta_2$ (red dots) respectively, using the Maia data. Phases are in radians

## *3.7 Squircle Data Versus Lifted Phase*

The word squircle as used in this report and in the sparse information provided about the microwave analyser equipment output refers to two separate ideas. One is the name *squircle* for the parallelogram shape with rounded corners that is seen when plotting raw phases against each other. We will from now on refer to this as *squircle shape*. Another is the name *squircle* given by the manufacturer to a column of data provided by the equipment alongside of the phase data. We will refer to this data as *squircle data*.

During the Study Group, we were unable to find out from the manufacturer or from the manual provided, what exactly the squircle data means or how it is calculated. So we tried to work this out by looking at the squircle data and comparing it to the lifted phase values we calculated from the raw phase data.

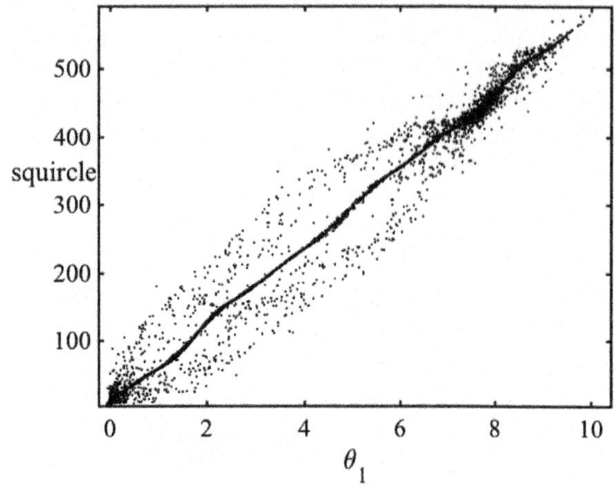


**Fig. 23** A plot of squircle data values versus our lifted phase values $\theta_1$, using the Maia data. Squircle data values are angles given in degrees, and lifted phase is in radians

Plots of the squircle data (taken directly from Maia data) versus lifted phase $\theta_1$ reveal that the squircle data does a similar thing to the lifted phase. These plots also confirm that squircle data is apparently based on treating the raw phase squircle shape as plotted in Fig. 5 like a clock face, and calculating the resulting angle (in degrees) that a clock hand would make with the initial phase position of that hand (when bauxite height is zero). This procedure is not as faithful to the raw phase data as the lifted phase process described here, as the clock face approach distorts the true phase, which is already an angle, by treating that angle as an arc length around the clock perimeter or squircle shape.

This may explain the wobble that is apparent in the plot in Fig. 23 of squircle data versus our lifted phase $\theta_1$. It is also apparent in the plot of squircle data versus bauxite height in Fig. 24. The main advantage of using our lifted phase over the squircle data provided by the microwave equipment is the removal of this wobble. On the scales plotted here, the wobble looks relatively small. However, in the operational height range of 250–350 mm, the wobble is more significant, as shown in the zoomed-in Fig. 25.

The ranges of values provided for squircle data, together with a comparison with our corresponding lifted phase values, suggest that the squircle data is provided in degrees rather than radians. We assume this to be true here.

## 3.8 Using Lifted Phase

The theoretical results presented later in Sect. 5 suggest that phase change and attenuation will be very close to linear when plotted against bauxite height or depth, at constant moisture content. Furthermore, Fig. 35 shows that the theoretical slope of attenuation versus depth plots decreases as moisture weight percentage increases,

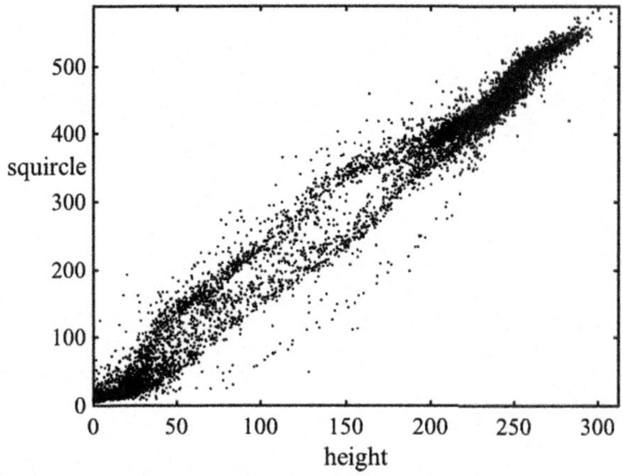


**Fig. 24** A plot of squircle data values (degrees) versus bauxite height (also called depth) in mm, using the Maia data

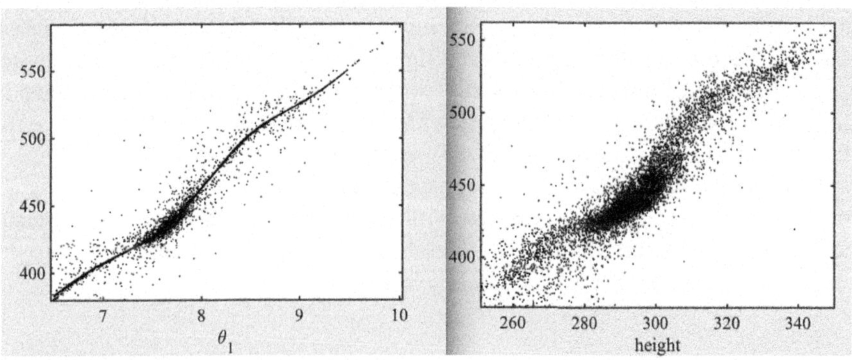


**Fig. 25** A zoomed version of the previous two plots of squircle data values versus lifted phase (radians) and bauxite height (also called depth) in mm, using the Maia data. The zoom-in focusses on bauxite depths of interest. The unlabelled vertical axis is squircle data (degrees) from the analyser

whilst the theoretical slope of phase versus depth plots increases with moisture content. It is also apparent in that figure that the attenuation is less sensitive to moisture content, as the attenuation lines bunch up as moisture increases through 9%. Note also that linearity in moisture content is not implied by Fig. 35. The spacing between the lines shows the effect of moisture. This analysis only considers the effect of moisture on the permittivity, and the dependence of the electrical conductivity $\sigma$ on moisture content is ignored.

The units associated with the attenuation data produced by the microwave analyser are not clear; we were advised that a value near zero corresponds to about 14 decibels of attenuation, and we observe from the data that zero attenuation (at zero height H), corresponds to a data value of 350. The data presented in Fig. 26 ranges from

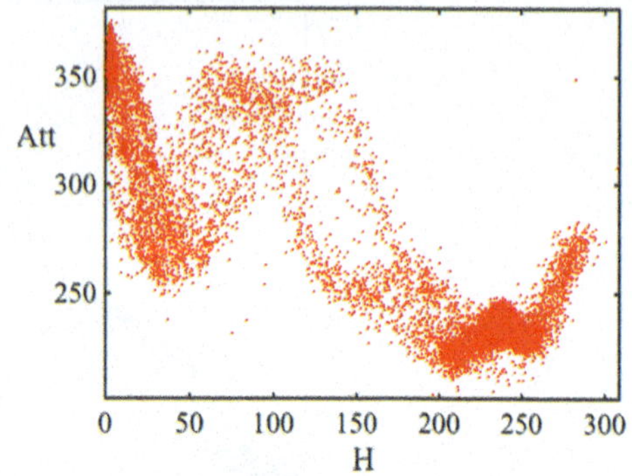


**Fig. 26** A plot of attenuation from the analyser data, versus bauxite height (mm), for the full Maia data set. The vertical scale is such that 350 corresponds to zero attenuation, and a zero value corresponds to 14 decibels of attenuation. See the text for a fuller discussion of the attenuation data scale

200 to 350 and indicates decreasing attenuation with increasing bauxite height, in contradiction to the theoretical results presented in Sect. 5. So there may be a sign reversal in the data. Furthermore, the attenuation data range would then correspond to about 6 Db for a height of 30 cm of bauxite, whereas the theoretical results indicate an attenuation of only 0.2 Db, about twenty times smaller than the data if the translation to Db is correct. Such details as units and signs for attenuation data become irrelevant if it is calibrated for regressing against, so we continue to investigate the shape of the attenuation data, and we persevere with the unknown units used to produce the data from the microwave analyser.

However, whilst the data for lifted phase plotted against bauxite height in Fig. 21 looks promising compared to the theory, the data for measured attenuation versus height as plotted in Fig. 26 does not look linear in general, in stark contrast to the plot in Fig. 21 of phase versus height.

## 3.9 *Lagged Height Measurement*

We investigated the appearance of two lines in plots like phase versus depth in Fig. 21, and realised that the time-behaviour of points was to come down one line as height drops rapidly, then go back up the other line as height goes back up rapidly. This led to trying different lags, of less than one second, by interpolating the height data which is measured adjacent to the microwave section. We found that a half a second delay removes the two lines, indicating that they are indeed due to the small time lag during rapid changes in height. The resulting plots of lifted phase versus height, and attenuation versus height, are in Fig. 27. Figure 28 also shows both lagged and unlagged phase versus height for a six-minute time period. The sawtooth original phase data also look much better with corrected height lags, as in Fig. 29.

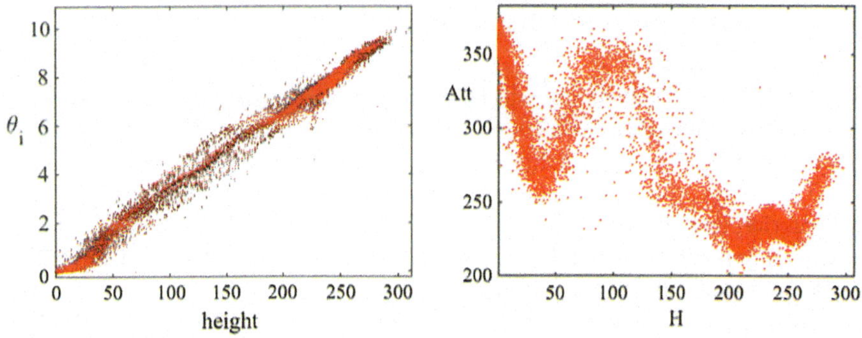


**Fig. 27** Plots of lifted phase (radians) and attenuation from the analyser data, versus bauxite height (mm), for the full Maia data set. The height data has been lagged by half a second, correcting the bilinear appearance seen before during times of rapid change in height. See the text for a discussion of the attenuation data scale

**Fig. 28** Plot of lifted phase in radians versus bauxite height in mm, for a short-time period in the Maia data set. The height data is plotted both lagged (by half a second, red dots) and unlagged (blue)

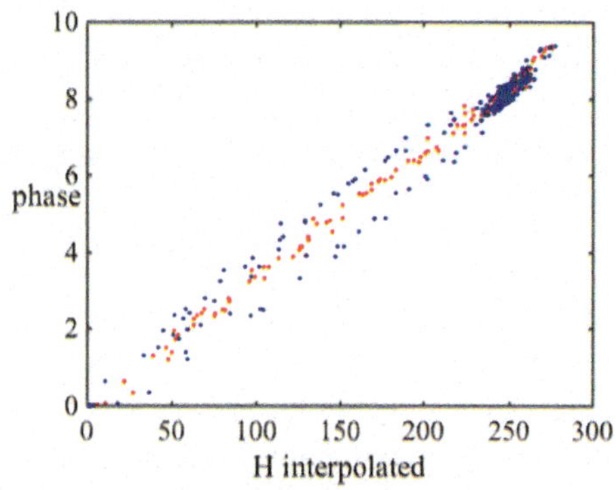


**Fig. 29** Plot of raw (unprojected) measured phases versus bauxite height, for the Maia data set. Height lag has been corrected, in contrast to Fig. 11 where the half-second lag in height measurement is not accounted for

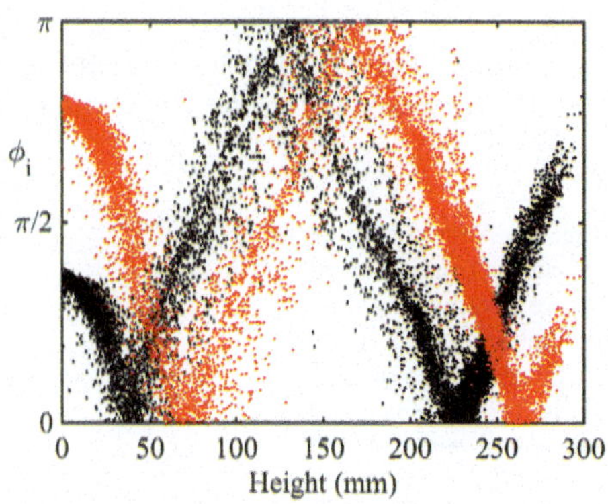

## 3.10 Attenuation and Interference

The behaviour of attenuation versus height is most likely due to internal reflection of microwaves at the interfaces between bauxite and air, with one wave coming directly through a thickness $H$ of bauxite, and a second, more attenuated and more lagged wave reflecting off the interfaces and coming through $2H$ more thickness of bauxite and interfering with the first, sometimes partially cancelling and sometimes partially augmenting it depending on the lag difference, which varies with bauxite height.

We assume the first wave is of the simple form

$$A_1 \sin(\omega t)\,,$$

where $A_1 = \exp(-k_2 H)$. Similarly, the second wave that has reflected once internally takes the form

$$A_2 \sin(\omega t + \phi)\,,$$

where $A_2 = \exp(-3k_2 H)$. Since the second wave goes through a total thickness of $3H$, then adding the above expressions and using simple trigonometry identities gives the resulting sinusoid as having an amplitude squared that is

$$A^2 = A_1^2 + A_2^2 + 2A_1 A_2 \cos(\phi)\,.$$

Using $\phi = 2k_1 H$, the phase shift of the second signal with respect to the first due to making two extra passes through the bauxite of depth $H$ with $k_1 = 3\pi/300$, the data shows a total phase change of $3\pi$ over a depth of 300mm. We plot the log of the above expression for $A^2$ to get the graph in Fig. 30.

We see from Fig. 30 that the locations of the minima are about right when compared with the data plotted in Fig. 27b, although the latter part of the value of attenuation is rather different for bauxite heights over 200mm. We also see that the dashed straight line that corresponds to the attenuation of the first signal without accounting for

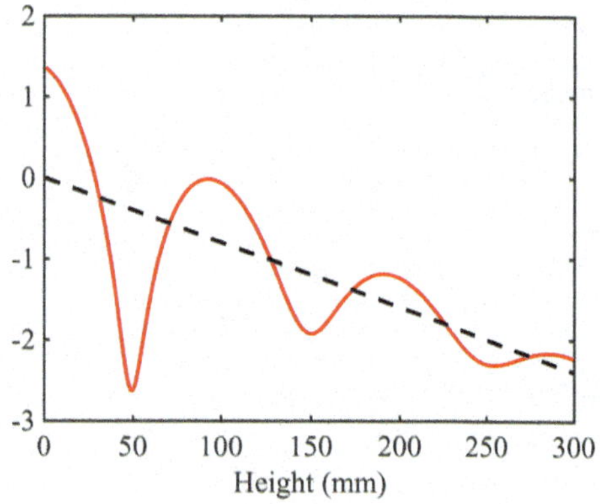


**Fig. 30** A plot of the log of $A^2$ versus bauxite height, where $A$ is the amplitude of two model signals summed together, and the second signal is a lagged and attenuated version of the first, as described in the text. The attenuation value used was $k_1 = 1/250$. The dashed line shows the log of the square of the amplitude of the first signal

interference effects looks similar to what could as easily be a fitted straight line through the curve representing the interfering sum of both signals.

## 3.11 Calibration of Analyser Data

The theoretical development presented later in Sect. 5 can be summarised as follows, with a view to calibrating the raw data provided by the analyser. The phase of the received signal is shifted from that of the reference signal by an amount $\phi$ which depends on bauxite height $H$ as in Eq. (24), where

$$\frac{\phi}{H} = A(\sqrt{\epsilon_r} - 1) = S_p \text{ (say)}. \tag{3}$$

The factor $A = 2\pi/\lambda_0$ depends on wavelength, whilst the relative permittivity $\epsilon_r$ is related to the moisture content $M$ by

$$\epsilon_r = aM + b .$$

The parameters $a$ and $b$ depend on bauxite properties and depend only weakly on porosity, and are implicitly given by Eq. (23b) in the theoretical development below. The attenuation $L$ also depends on $H$ and $\epsilon_r$ as in Eq. (25) giving

$$\frac{L}{H} = \frac{c}{\sqrt{\epsilon_r}} = S_a \text{ (say)},$$

provided we ignore reflection and interference effects, and the possibility that the conductance that appears in $c$ also depends on moisture content.

That is, both phase and attenuation depend (differently) on permittivity, and hence on moisture. A useful way to reduce noise might be to plot the lifted phase $\theta$ versus height, and attenuation versus height, and regress to get slopes $S_p = A(\sqrt{\epsilon_r} - 1)$ for phase and $S_a = \frac{c}{\sqrt{\epsilon_r}}$ for attenuation. But in the same spirit as the approach used in McKeown et al. (2017), where the density varies but the thickness is constant, whereas in our case bauxite density does not vary much but the height does vary significantly, we resist the temptation to regress just yet. Instead we eliminate $\epsilon_r$ to obtain

$$\frac{\theta}{H} = A\left(c\frac{H}{L} - 1\right) .$$

This indicates that if we regress $\theta/H$ against $H/L$, we get a line with slope $Ac$ and intercept $-A$. The regression should be one that acknowledges that there is noise in both variables; that is, errors-in-variables or Deming regression. When using Deming regression, it is normal practice to scale so that both variables have similar mean and variances. Then after regression we have estimates for values of $A$ and $c$.

Now we can calibrate using some short-time windows centred on times when moisture is measured in the lab. Note that the above equations rearrange to give

$$M = \alpha \left(1 + \frac{S_p}{A}\right)^2 + \beta,$$

where $\alpha$ and $\beta$ depend on $a$ and $b$, hence on bauxite properties, and $S_p$ is the slope of a Deming regression of $\theta$ against $H$. Thus, if we have several lab measurements of moisture $M$ at different times, we can (Deming) regress them against the corresponding values of the moisture measure

$$\text{M data} = (1 + S_p/A)^2$$

to find $\alpha$ and $\beta$.

An example of a short-time window and the resulting regressions is shown in Fig. 31. They look more convincing when height is limited to values greater than 50 mm, as illustrated. Attenuation has been altered from the value output by the analyser by taking the difference: maximum attenuation minus the value of attenuation, so that the resulting value increases from zero as height increases. We suspect the attenuation should not be used because of the large effect we are seeing from internal reflections and the resulting interference as height changes, which are not in the model.

It is important when calibrating using lab moisture values that the sample time matches the analyser time. We were advised that the times provided for lab moisture measurements lag analyser values by about ten minutes. We refine a single lag time by plotting a time series of 300 6 min wide windows of phase slope values $S_p$, evenly spaced over a period of time that contains the hour of lab measurements on the same graph as the lab moisture measurements. We first adjust the phase slope values to have the same mean and variance as the lab measurements, so that they show up on the same plotted scale. We found that lagging the phase slope values by ten and a half minutes gives a good match as shown in Fig. 32.

The combination method does not look convincing. As noted in the conclusions, the theory that this method is based on ignores the effect of moisture content on conductivity, and internal reflections are not modelled, rendering any method that relies strongly on attenuation suspect.

However, the slopes $S_p$ of phase versus height do look useful as measures of moisture content, as evidenced in Fig. 33 where lab moistures are plotted against rescaled values $S'_p$.

The theoretical dependence of phase on moisture content is not linear, as is evident from the spacing of the lines in Fig. 35a at constant bauxite height, and from the dependence of phase on the square root of $\epsilon_r$ in Eq. (3), and hence on the square root of $aM + b$ with constant $a$ and $b$.

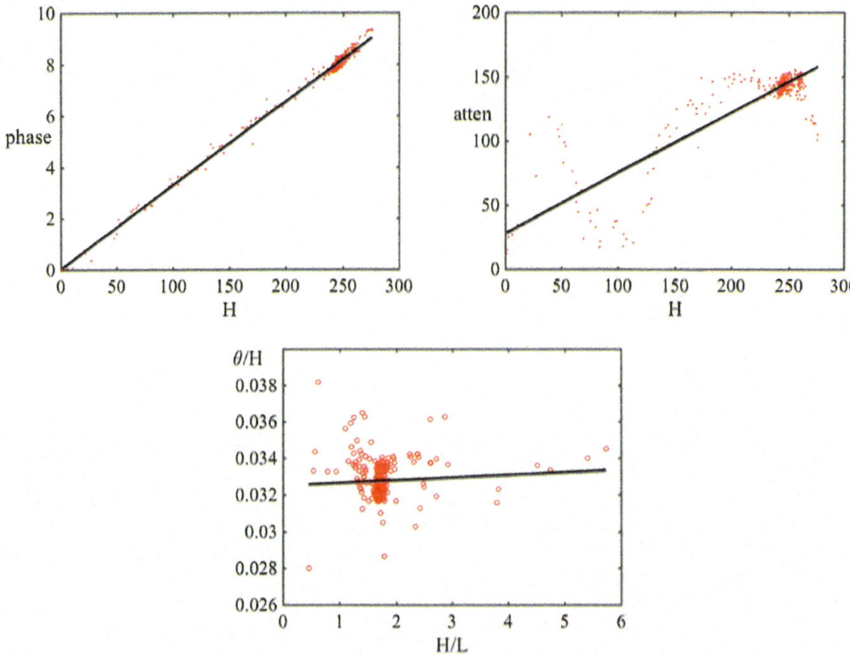


**Fig. 31** An example of a short-time window on Maia data, and the resulting regressions of lifted phase versus height, attenuation vs height, and phase over height vs height over attenuation for heights greater than 50 mm. Phases are in radians, heights are in mm. Attenuation units are discussed in the second paragraph of Sect. 3.8. Note that the slope of lifted phase vs height, 0.033 from the regression shown in the first plot, may also be observed as the ordinate value in the last plot and in that plot it is apparent that this number is almost independent of attenuation (and height). It is the strongest regression variable

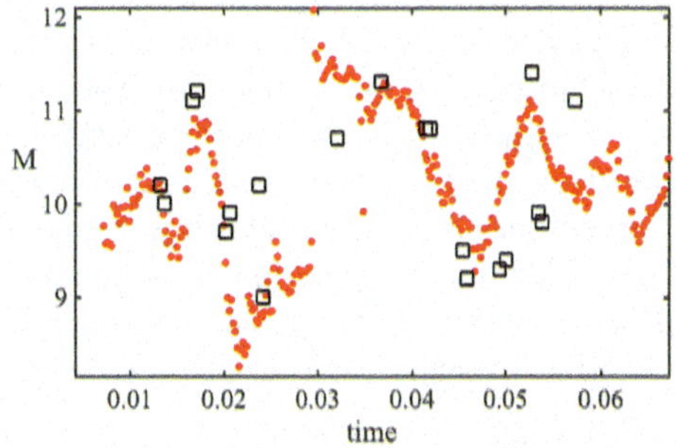


**Fig. 32** Lab moisture values (weight percent, wet basis) (black squares) for Maia plotted against time (decimal days), on the same graphs as values of phase slopes $S_p$ (red dots). The phase slopes have been lagged by ten and a half minutes to get the match seen here. Prior to lagging, phase slopes were rescaled to have the same mean and variance as the moisture values, to help the hand-matching that led to this lag

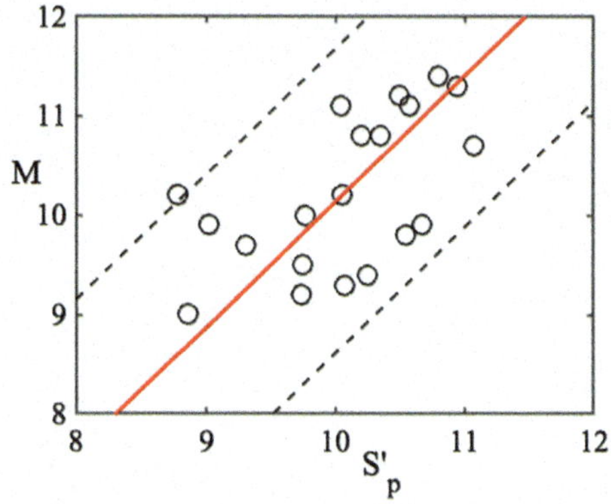


**Fig. 33** Lab moisture values M (weight percent, wet basis) for Maia plotted against rescaled phase slopes $S'_p$ calculated at the same times from analyser data, after correcting the lab times to sample time, and after rescaling phase slopes $S_p$ to values $S'_p$ with mean and variance that match those of lab moistures. The data are shown using symbols, and the solid line is a fitted Deming regression line. The dashed lines are an indication of 95% confidence values of the regression, obtained by calculating the standard (root mean square) error of the regression, and adding and subtracting twice this standard error to the fitted line

However, for small changes in moisture content, in particular for the ranges of moisture content encountered in the data under consideration in this paper, a linear approximation may be adequate. The resulting relationship between lab moisture values $M$ and (rescaled) slopes $S'_p$ is shown in Fig. 33.

We are interested in testing linear relationships also because the analyser equipment is known to assume such linear relationships when it performs its own regressions to obtain values for moisture content.

Using the corrected lab moisture sampling times, slopes of phase angle versus bauxite height were obtained in six-minute windows centred on all twenty lab times. The slope values were adjusted by a linear transformation to have the same mean and variance as the lab moistures, then Deming regressed against them. The slope and intercept parameters from that regression were used to generate the solid straight line in Fig. 33. The dashed lines on either side result from adding and subtracting twice the standard error of the fitted straight line, estimated by calculating the root mean squared residual over all twenty points and provide an estimate of the 95% confidence prediction interval for the value of predicted individual moisture data points.

The standard error for the fitted Deming regression line is about 0.77, indicating a 95% confidence interval on individual points of $\pm 1.5$ weight percent moisture content. This is a little larger than the $\pm 1\%$ RAAL were hoping for.

# 4 Polarisation Theory and Phase Shift

## 4.1 Waves in Materials

The fundamental physical principle behind the microwave moisture detector is the large dipole moment of a water molecule as compared to other compounds typically found in bauxite. Before considering the ramifications of this we begin by considering Maxwell's equations for electromagnetic waves in a dielectric material

$$\nabla \cdot \mathbf{D} = \rho_f, \qquad \nabla \cdot \mathbf{B} = 0, \qquad \nabla \times \mathbf{E} = -\frac{\partial \mathbf{B}}{\partial t}, \qquad \nabla \times \mathbf{H} = \mathbf{J}_f + \frac{\partial \mathbf{D}}{\partial t}, \tag{4}$$

where $\mathbf{E}$ is the electric field intensity, $\mathbf{D}$ the electric displacement, $\mathbf{H}$ the magnetic field intensity, $\mathbf{B}$ the magnetic induction, $\mathbf{J}_f$ the current density and $\rho_f$ the charge density. Notice that there is an internal consistency in the expressions since

$$\nabla \cdot \nabla \times \mathbf{E} = -\frac{\partial}{\partial t}\nabla \cdot B = 0, \qquad \nabla \cdot \nabla \times \mathbf{H} = \nabla \cdot \mathbf{J}_f + \frac{\partial \rho_f}{\partial t} = 0,$$

the latter implying that charge is conserved. When an electromagnetic wave passes through a material its response to electric and magnetic fields depends on the atomic structure of the material. For an electric field the displacement is $\mathbf{D} = \epsilon_0 \mathbf{E} + \mathbf{P}$ where $P$ is the polarisation field of the dielectric. Assuming this polarisation is in the same direction as $\mathbf{E}$ and proportional to it, so that $\mathbf{P} = \epsilon_0 \chi_e \mathbf{E}$ then $\mathbf{D} = \epsilon_0(1 + \chi_e)\mathbf{E} = \epsilon \mathbf{E}$. Correspondingly, the magnetic induction is $\mathbf{B} = \mu_0(\mathbf{H} + \mathbf{M})$, where $\mathbf{M}$ is a magnetization field of the dielectric. Assuming, as with the electric field that this induced field is proportional to the field that generated it, $\mathbf{M} = \mu_0 \chi_m \mathbf{H}$ and $\mathbf{B} = \mu_0(1 + \chi_m)\mathbf{H} = \mu \mathbf{H}$.[1] If we also assume that the material is conductive with a scalar conductivity of $\sigma$ so that $\mathbf{J}_f = \sigma \mathbf{E}$ then from (4)

$$-\nabla \times \nabla \times \mathbf{E} = \frac{\partial}{\partial t}\nabla \times \mathbf{B} = \mu\sigma \frac{\partial \mathbf{E}}{\partial t} + \mu\epsilon \frac{\partial^2 \mathbf{E}}{\partial t^2} = -\nabla(\nabla \cdot \mathbf{E}) + \nabla^2 \mathbf{E} = \nabla^2 \mathbf{E}, \tag{5a}$$

$$\begin{aligned} -\nabla \times \nabla \times \mathbf{H} &= -\nabla \times \left(\sigma \mathbf{E} + \epsilon \frac{\partial \mathbf{E}}{\partial t}\right) = \mu\sigma \frac{\partial \mathbf{H}}{\partial t} + \mu\epsilon \frac{\partial^2 \mathbf{B}}{\partial t^2} \\ &= -\nabla(\nabla \cdot \mathbf{H}) + \nabla^2 \mathbf{H} = \nabla^2 \mathbf{H}, \end{aligned} \tag{5b}$$

---

[1] When propagating through a vacuum, the permittivity of free space is simply $\epsilon_0 = 8.85 \times 10^{-12}$ $\mathrm{F m^{-1}}$ and the permeability of free space is $\mu_0 = 4\pi \times 10^{-7}$ $\mathrm{H m^{-1}}$ which corresponds to a propagation speed of $c = (\epsilon_0 \mu_0)^{-1/2}$ $\mathrm{m s^{-1}}$. The quantities $\chi_e$ and $\chi_m$ are known as the electric and magnetic susceptibilities of the material and are typically tensors with structure that reflects the underlying symmetries of the atomic lattice of the material. For an isotropic, noncrystalline material these parameters are taken as scalars.

illustrating that the electric and magnetic fields satisfy the same damped wave equation.

For convenience, consider a monochromatic electric field with angular frequency $\omega$ making excursions in the $\hat{\mathbf{e}}_y$ direction and propagating in the $\hat{\mathbf{e}}_x$ direction.[2] The corresponding $\mathbf{H}$ field must satisfy $\nabla \times \mathbf{E} = -\mu\partial\mathbf{H}/\partial t$ and therefore

$$\mathbf{E}(t,\mathbf{x}) = E_0 \mathrm{e}^{ikx-i\omega t}\hat{\mathbf{e}}_y, \qquad \mathbf{H}(t,\mathbf{x}) = \frac{k}{\mu\omega} E_0 \mathrm{e}^{ikx-i\omega t}\hat{\mathbf{e}}_z . \tag{6}$$

Substituting these forms into the differential equation (5a) leads to the wavenumber[3] $k$ being given by

$$k^2 = \mu\epsilon\omega^2 + i\mu\sigma\omega . \tag{7}$$

We treat $k$ as a complex value $k = k_a + ik_b$ under the assumption that both $\epsilon$ and $\sigma$ are real valued. This results in

$$k_a = \omega\left(\frac{\mu\epsilon}{2}\right)^{1/2}\left(1+\sqrt{1+\left(\frac{\sigma}{\epsilon\omega}\right)^2}\right)^{1/2} = \omega\,(\mu\epsilon)^{1/2}\left(1+\frac{1}{8}\left(\frac{\sigma}{\epsilon\omega}\right)^2 - \frac{5}{128}\left(\frac{\sigma}{\epsilon\omega}\right)^4 + \cdots\right), \tag{8a}$$

$$k_b = \sigma\left(\frac{\mu}{2\epsilon}\right)^{1/2}\left(1+\sqrt{1+\left(\frac{\sigma}{\epsilon\omega}\right)^2}\right)^{-1/2} = \frac{\sigma}{2}\left(\frac{\mu}{\epsilon}\right)^{1/2}\left(1-\frac{1}{8}\left(\frac{\sigma}{\epsilon\omega}\right)^2 + \frac{7}{128}\left(\frac{\sigma}{\epsilon\omega}\right)^4 + \cdots\right), \tag{8b}$$

choosing the branch cut to match with the limit of $k \to k_a$ as $\sigma \to 0$.

A popular choice in studies of microwave propagation through grains (for example, Nelson and Trabelsi (2010)) is to define the permittivity to be complex, following Lorrain et al. (1987)), and Landau and Lifshitz (1984), and to do away with explicit references to conductivity. With our choice of $e^{-i\omega t}$ for the time dependence we write

$$k^2 = \mu\epsilon\omega^2 + i\mu\sigma\omega \tag{9}$$

$$= \mu\omega^2\left(\epsilon + i\frac{\sigma}{\omega}\right) \tag{10}$$

$$= \mu\omega^2\epsilon_c, \tag{11}$$

where the complex-valued permittivity is

[2] The propagation direction is given by the wavenumber that satisfies $\mathbf{k}\cdot\mathbf{x} = (k,0,0)\cdot(x,y,z) = kx$.

[3] The angular $\omega$ and linear $f$ frequencies are related by $\omega = 2\pi f$. The wavelength $\lambda$ depends on the phase speed of medium so that $\lambda f = v = c/n$ where $n$ is the refractive index. The wavenumber $k$ is the spatial frequency of the wave measured in either cycles per unit of distance or radians per unit distance. Choosing the latter, $k\lambda = 2\pi$. For a complex-valued wavenumber the real part corresponds to the spatial frequency whilst the imaginary part describes the attenuation.

$$\epsilon_c = \epsilon' + i\epsilon'' = \epsilon + i\frac{\sigma}{\omega}. \tag{12}$$

Then it is clear that where a paper refers to $\epsilon'$, this is equal to our real value for permittivity, and that $\epsilon'' = \frac{\sigma}{\omega}$.

An upper bound on the conductivity $\sigma$ of the material to apply the asymptotic series gives $\epsilon\omega > \epsilon_0\omega = 50$ mS m$^{-1}$ at $f = 900$ MHz. For comparison, the conductivity of sea water is about 5000 mS m$^{-1}$ and damp wood has a conductivity of about 0.1 mS m$^{-1}$. Of particular convenience for studying the passage of electromagnetism through matter is the index of refraction $n$ and the absorption coefficient $\alpha$ defined as

$$n = \frac{c}{v} = c(\mu\epsilon)^{1/2}, \qquad \alpha = \sigma\left(\frac{\mu}{\epsilon}\right)^{1/2}. \tag{13}$$

From these expressions, we expect that the microwave radiation incident on a section of bauxite suffers both a phase shift and an attenuation which are functions of the material parameters. Rather than using the quantities $\epsilon$ and $\mu$ directly, we use the relative versions defined as $\epsilon = \epsilon_0\epsilon_r$, $\mu = \mu_0\mu_r$ so that with respect to these quantities,

$$n = \left(\frac{\mu_0\mu_r\epsilon_0\epsilon_r}{\mu_0\epsilon_0}\right)^{1/2} \simeq \epsilon_r^{1/2}, \quad \alpha = \sigma\left(\frac{\mu}{\epsilon}\right)^{1/2} = \sigma\left(\frac{\mu_0\mu_r}{\epsilon_0\epsilon_r}\right)^{1/2} \simeq \frac{\sigma Z_0}{\epsilon_r^{1/2}}, \tag{14}$$

where $Z_0 = (\mu_0/\epsilon_0)^{1/2} = 376.73\Omega$ is the intrinsic impedance of free space and setting $\mu_r \simeq 1$ because the varying list of materials that make up bauxite are assumed to be typically nonmagnetic (i.e. without steel, nickel or iron components). With these final definitions,

$$k_a = \frac{2\pi}{\lambda_0}\left(\frac{\epsilon_r}{2}\right)^{1/2}\left(1+\sqrt{1+\left(\frac{\sigma}{\epsilon_0\epsilon_r\omega}\right)^2}\right)^{1/2}, \quad k_b = \frac{\sigma Z_0}{(2\epsilon_r)^{1/2}}\left(1+\sqrt{1+\left(\frac{\sigma}{\epsilon_0\epsilon_r\omega}\right)^2}\right)^{-1/2}, \tag{15}$$

$\lambda_0 = f/c$ the wavelength of the radiation in free space and with the electromagnetic fields taking the form

$$\mathbf{E}(t,\mathbf{x}) = E_0 e^{-k_b x} e^{ik_a x - i\omega t}\hat{\mathbf{e}}_y, \quad \mathbf{H}(t,\mathbf{x}) = \frac{|k|}{\mu_0\omega} E_0 e^{i\delta} e^{-k_b x} e^{ik_a x - i\omega t}\hat{\mathbf{e}}_z, \quad |\mathbf{S}(\mathbf{x})| = \frac{|k|}{\mu_0\omega} E_0^2 e^{-2k_b x}\hat{\mathbf{e}}_x, \tag{16}$$

where $\mathbf{S} = \mathbf{E} \times \mathbf{H}$ is the Poynting vector giving the flow of energy density, and $k = |k|e^{i\delta}$ defines a phase shift between the electric and magnetic fields as they pass through the medium.

The nondimensional parameter $\sigma/\epsilon\omega$ hints at a useful analogy for the dielectric material. Consider a length of bauxite of cross sectional area $A$ and length $\ell$ with conductivity $\sigma$ and permittivity $\epsilon$. The electrical resistance of this block of material is $R = \ell/A\sigma$ whilst the capacitance is $C = \epsilon A/\ell$. This shows that $\sigma/\epsilon\omega$ is simply the ratio of period of the applied frequency to the $RC$ time constant of the medium. Moreover,

$$\frac{k_b}{k_a} = \frac{\frac{\sigma}{\epsilon\omega}}{1+\sqrt{1+(\frac{\sigma}{\epsilon\omega})^2}} \rightarrow \begin{cases} \frac{\sigma}{2\epsilon\omega} + \mathcal{O}((\frac{\sigma}{\epsilon\omega})^3), & \frac{\sigma}{\epsilon\omega} \rightarrow 0, \\ 1 + \mathcal{O}((\frac{\sigma}{\epsilon\omega})^{-1}), & \frac{\sigma}{2\epsilon\omega} \rightarrow \infty \end{cases}$$

implying that for low conductivity $\delta \sim 0$ whilst $\delta \rightarrow \pi/4$ for a highly conductive medium. Also of importance is the reference electromagnetic signals that do not pass through the material. In this case one sets $\epsilon_r = 1$ and $\sigma = 0$ to find that $k = (\epsilon_0\mu_0)^{1/2}\omega = \omega/c = 2\pi/\lambda_0$ and

$$\mathbf{E}_{\text{ref}}(t, \mathbf{x}) = E_0 e^{2\pi i x/\lambda_0 - i\omega t}\hat{\mathbf{e}}_y, \quad \mathbf{H}_{\text{ref}}(t, \mathbf{x}) = \frac{1}{Z_0} E_0 e^{2\pi i x/\lambda_0 - i\omega t}\hat{\mathbf{e}}_z, \quad |\mathbf{S}_{\text{ref}}(\mathbf{x})| = \frac{1}{Z_0} E_0^2 \hat{\mathbf{e}}_x. \tag{17}$$

## 4.2 Dielectric Polarisation

When a dielectric experiences a macroscopic external electric field $\mathbf{E}$, it responds by generating a dielectric polarisation $\mathbf{P}$, defined as the dipole moment per unit volume. The dielectric susceptibility $\chi_e = \epsilon_r - 1$ connects these two quantities so that $\mathbf{P} = \epsilon_0\chi_e\mathbf{E}$ assuming that $\mathbf{E}$ and $\mathbf{P}$ remain aligned.[4] The local field at any particular atom is the sum of the electric field from external sources and of the field from the dipoles within the specimen. If one assumes a spherically shaped inclusion then this local field takes the form

$$\mathbf{E}_{\text{local}} = \mathbf{E} + \frac{1}{3\epsilon_0}\mathbf{P}. \tag{18}$$

The factor of $1/3$ reflects the spherical symmetry and for other shapes the exact factor will vary but still be of order one.

The local field that each molecule experiences due to this effective local field stretches the dipoles, thereby generating a dipole moment. For each type of molecule this effect is encoded in the polarisability, $\alpha$, defined in terms of the local electric field at the molecule as $\mathbf{p} = \alpha\mathbf{E}_{\text{local}}$. Each separate molecule has its own characteristic polarisability from the local electric field that it experiences implying that the total polarisation is $\mathbf{P} = \sum_j N_j\mathbf{p}_j$ where $N_j$ is the concentration of species $j$. In terms of the polarisabilities of the atoms and the local electric field, the polarisation of the material may be expressed approximately as

[4] The polarisation $\mathbf{P}$ of a material is related to the applied electric field $\mathbf{E}$ through a rank two electric susceptibility tensor $\boldsymbol{\chi}_e$ that reflects the underlying atomic structure. For bulk bauxite a scalar quantity was deemed sufficient due to the random orientation of the material on the conveyor.

$$\epsilon_0(\epsilon_r - 1)\mathbf{E} = \left(\mathbf{E} + \frac{1}{3}(\epsilon_r - 1)\mathbf{E}\right)\sum_j N_j\alpha_j\,, \tag{19}$$

by assuming that the local field is the same for all of the molecules. Rearranging gives the Clausius-Mossotti equation (hereafter referred to as the C-M equation),

$$\frac{\epsilon_r - 1}{\epsilon_r + 2} = \frac{1}{3\epsilon_0}\sum_j N_j\alpha_j. \tag{20}$$

The crucial point here is that the right-hand side of (20) is additive with respect to the number of polarisers. We consider the bauxite as being composed of three separate classifications of material. Namely, water, air and solids with the term solids being used rather loosely and referring to all of the remaining components in the mineral, irrespective of their state. Let $p$ be the mole fraction of water, that is, the number of moles of water per mole of mineral. Correspondingly $q$ denotes the mole fraction of air and $1 - p - q$ is the mole fraction of the remainder consists primarily of solids. Let each of these components have respective permeabilities of $\epsilon_w \simeq 80.1$, $\epsilon_a \simeq 1.0$ and $2 \leq \epsilon_s \leq 10$. According to (20) the relative permeability of this mixture would satisfy

$$\frac{\epsilon_r - 1}{\epsilon_r + 2} = p\frac{\epsilon_w - 1}{\epsilon_w + 2} + (1 - p - q)\frac{\epsilon_s - 1}{\epsilon_s + 2} + q\frac{\epsilon_a - 1}{\epsilon_a + 2}. \tag{21}$$

Connected to this is the weight percent of the water within the bauxite given by

$$\frac{1}{100}M = \frac{\text{weight of water}}{\text{weight of wet bauxite}} = \frac{pM_w}{pM_w + (1 - p - q)M_s + qM_a}, \tag{22}$$

where the $M_j$ are the molecular weights given in units of grams per mole by $M_w \simeq 18.02$, $M_a \simeq 28.97$, $110 \leq M_s \leq 140$. Note that this assumes that the mineral is weighed in a vacuum since it neglects the small buoyancy force experienced by the displaced air of the mineral sample.

One of the significant challenges is the variability of the solid matrix that characterises the bauxite seen in the parameters $2 \leq \epsilon_s \leq 20$ and $110 \leq M_s \leq 140$. Any calibration efforts to measure the moisture must embrace and systematically compensate for these variations. Solving (21)–(22) to find $\epsilon_r$ as a function of the moisture we obtain

$$\epsilon_r = \frac{(100 - M)M_w(\epsilon_w + 2)(3\epsilon_s - 2q(\epsilon_s - 1)) + MM_s(\epsilon_s + 2)(3\epsilon_w - 2q(\epsilon_w - 1)) + 6MM_aq(\epsilon_w - \epsilon_s)}{(100 - M)M_w(\epsilon_w + 2)(3 + q(\epsilon_s - 1)) + MM_s(\epsilon_s + 2)(3 + q(\epsilon_w - 1)) - 3MM_aq(\epsilon_w - \epsilon_s)} \tag{23a}$$

$$= \frac{3\epsilon_s - 2q(\epsilon_s - 1)}{3 + q(\epsilon_s - 1)} + \frac{(\epsilon_s + 2)(\epsilon_w - \epsilon_s)(qM_a + (1 - q)M_s)}{M_w(\epsilon_w + 2)(1 + \frac{q}{3}(\epsilon_s - 1))^2}\frac{M}{100} + \mathcal{O}(M^2). \tag{23b}$$

This expression reveals that the variation in permittivity at zero moisture is due to the variability in $\epsilon_s$ and independent of $M_s$ with the air content of the mineral playing

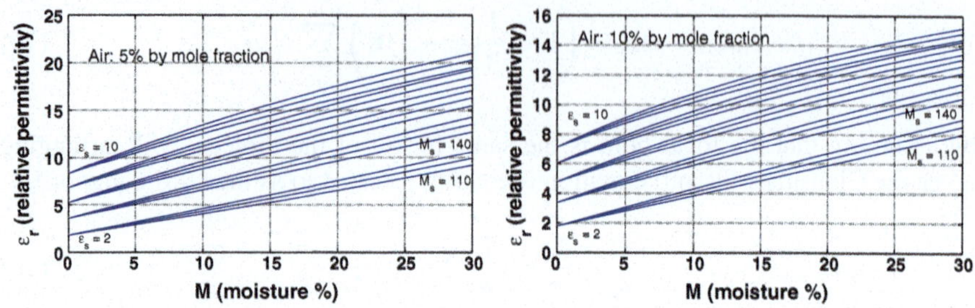


**Fig. 34** The variation of the predicted electric permittivity as a function of weight percent of moisture for nominal values of $M_w = 18.02$, $M_a = 28.97$, $M_s \in \{110, 125, 140\}$, all in units of grams per mole, $\epsilon_w = 80.1$, $\epsilon_a = 1.0$ and $\epsilon_s \in \{2, 4, 6, 8, 10\}$. To the left $q = 5\%$ air mole fraction and to the right $q = 10\%$ air mole fraction

only a weak role. The coefficient in the linear term details the sensitivity of $\epsilon_r$ with respect to changes in moisture and reveals that the factor $\epsilon_w - \epsilon_s$ is crucial and this is modulated with a factor of an effective size of $M_s(\epsilon_s + 2)/M_w(\epsilon_w + 2)$. As with the zero point level, the air content plays a weak role. Figure 34 illustrates the behaviour of the predicted $\epsilon_r$ as a function of the moisture for mineral samples with 5% and 10% air mole fractions. At 5% air content $\Delta\epsilon_r$ is approximately 10, irrespective of the $M_s$ and $\epsilon_s$ and at 10% air, $\Delta\epsilon_r$ reduces to about 6, again irrespective of $M_s$ and $\epsilon_s$.

## 5 Theory of Operation

In its most basic form, the moisture detector operates by detecting changes in the received phase and attenuation of a microwave burst that is sent through a sample of bauxite moving on a conveyor into the refinery. The dielectric properties and in particular the dielectric permittivity depend on the amount of moisture in the bauxite and on the varying properties of bauxite from ship to ship. Particular to bauxite is the challenge that it does not have a specific composition, but it is rich in aluminium minerals oxides, clays and a variety of other insoluble materials. Because of this, its effective molecular weight and relative permittivity may change significantly from one sample to the next.

Three basic principles connect to explain the behaviour of the moisture detector.

- The composition of the bauxite (minerals, water, air) determine an effective permeability given by the mole fraction of the components (C-M equation).
- The weight percent of water is a function of the mole fraction of the various components weighted by their molecular weights.
- The phase shift of the electromagnetic wave passing through the material is proportional to the index of refraction of the material which is also determined by this effective permeability.

The first two points give expression (23) relating the moisture level to the electric permittivity and for the phase shift, Eqs. (15), (16) and (17) combine to give

$$\Delta\phi = \frac{2\pi H}{\lambda_0}\left(\left(\frac{\epsilon_r}{2}\right)^{1/2}\left(1+\sqrt{1+\left(\frac{\sigma}{\epsilon_0\epsilon_r\omega}\right)^2}\right)^{1/2} - 1\right) \tag{24a}$$

$$= \frac{2\pi H}{\lambda_0}\left(\epsilon_r^{1/2} - 1 + \frac{1}{8}\left(\frac{\sigma}{\epsilon_0\omega}\right)^2 \epsilon_r^{-3/2} + \mathcal{O}\left(\left(\frac{\sigma}{\epsilon_0\omega}\right)^4\right)\right) \tag{24b}$$

for a propagation distance of length $H$. The attenuation in decibels can be found using the same set of expressions as

$$L_{\text{dB}} = -20\log_{10}\left(\frac{|\mathbf{E}|}{|\mathbf{E}_{\text{ref}}|}\right) = 8.686\frac{\sigma Z_0 H}{(2\epsilon_r)^{1/2}}\left(1+\sqrt{1+\left(\frac{\sigma}{\epsilon_0\epsilon_r\omega}\right)^2}\right)^{-1/2} \tag{25a}$$

$$= 4.343\sigma Z_0 H\left(\epsilon_r^{-1/2} - \frac{1}{8}\left(\frac{\sigma}{\epsilon_0\omega}\right)^2 \epsilon_r^{-5/2} + \mathcal{O}\left(\left(\frac{\sigma}{\epsilon_0\omega}\right)^4\right)\right). \tag{25b}$$

The expected idealised behaviour of the phase shift and loss as a function of depth for a variety of moistures comes from combining (23), (24) and (25) with the results shown in Fig. 35. We have taken the same nominal parameter values as in the previous figure and specified that $M_s = 110$, $\epsilon_s = 5$, $q = 5\%$ air mole fraction and a medium conductivity of $\sigma = 1\ \text{mS}\,\text{m}^{-1}$ at $f = 900$ MHz.

## 6 Attenuation and the Beer-Lambert Law

The Beer-Lambert law for attenuation of an electromagnetic wave is usually used to determine the concentration of a chemical dissolved in a solution. It is a particle approach to radiation rather than a wave approach, and it assumes that radiation is scattered off scattering centres in the medium, forcing it to travel further than without scattering. It also assumes there is some constant attenuation rate with total distance travelled. The result is an exponential decay of intensity with thickness of material. Two things are doubtful about using this law here: the radiation should travel much further than one wavelength since it can only detect or scatter off objects larger than its wavelength, and the incident flux must not influence the atoms or molecules being detected—it should act as a non-invasive probe of the species under study. Both of these conditions are clearly not obeyed by our moist bauxite. It has a thickness which is less than or close to the wavelength of the incident microwaves, and it is a polarising medium, with dipolar water molecules and various ions which do interact with the incident microwaves.

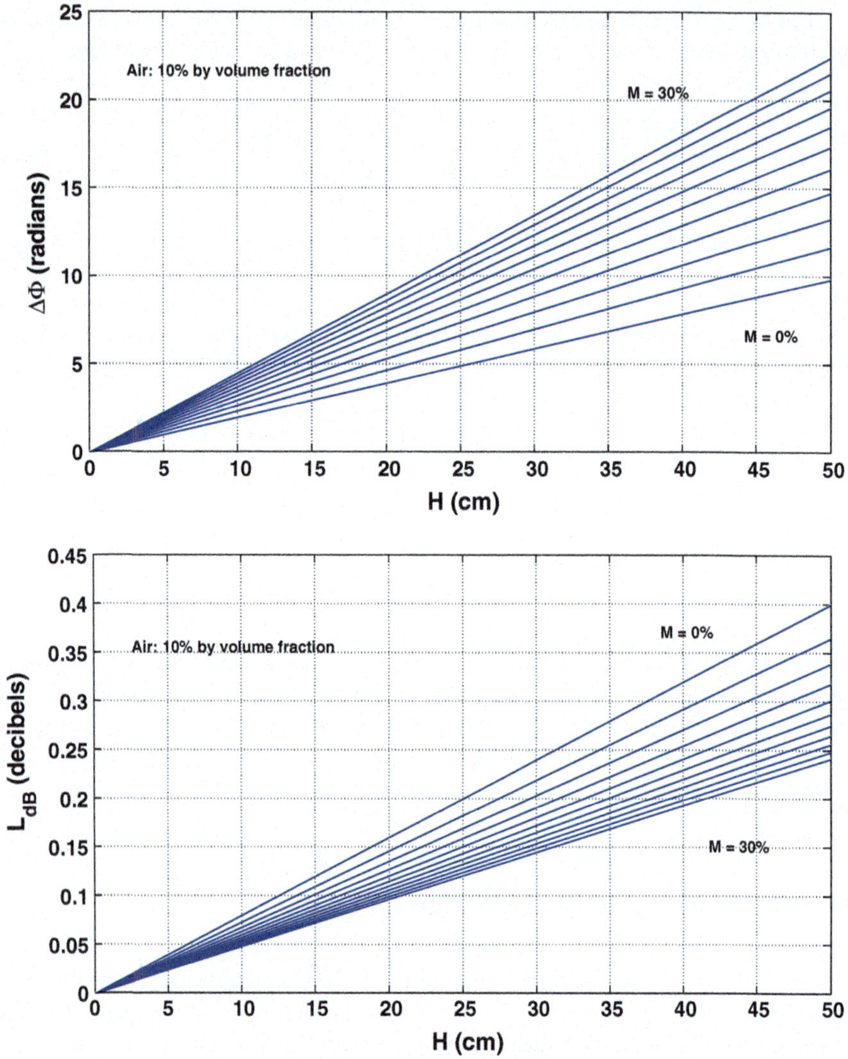


**Fig. 35** The idealised behaviour of the phase shift $\Delta\phi$ and the loss $L_{\text{dB}}$ as a function of depth $H$ for a variety of evenly spaced moisture levels $0 \leq M \leq 30\%$. Nominal values used are $M_w = 18.02$, $M_a = 28.97$, $M_s = 110$, all in units of grams per mole, $\epsilon_w = 80.1$, $\epsilon_a = 1.0$ and $\epsilon_s = 5$, $q = 10\%$ air mole fraction, $\sigma = 1$ mS.m$^{-1}$, $\lambda_0 = 0.333$ m and $\omega = 2\pi f$ with $f = 900$ MHz

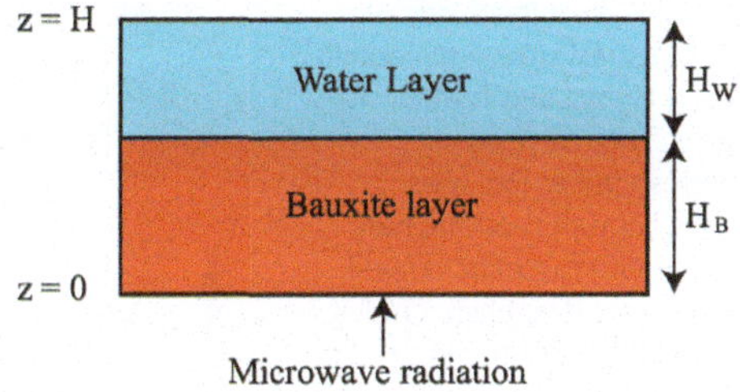


**Fig. 36** Bauxite and water as two independent horizontal layers

Nevertheless, it turns out that there is some support for this relatively simple scattering attenuation model, in the data. We model the medium as consecutive layers of water and dry bauxite, as illustrated in Fig. 36.

The strength of the emitter signal decreases as it passes through the bauxite and the water. This is due to the energy of the signal being absorbed as it passes through the media. In particular, this decrease in signal strength is proportional to its magnitude. This relationship can be posed mathematically as

$$\frac{\mathrm{d}P}{\mathrm{d}z} = -K_e P, \tag{26}$$

where $P$ is the strength of the signal (in decibels), $z$ is the distance in the vertical direction (with $z = 0$ and $z = H$ located at the top and bottom, respectively, of the media in the conveyor), and $K_e$ is the extinction coefficient, which is proportionality constant. The signal strength remaining after a microwave has passed through the bauxite and water is

$$P(H) = P_0 \exp\left(-K_{e,B} H_B - K_{e,W} H_W\right), \tag{27}$$

where $P_0$ is the signal strength at $z = 0$, and $K_{e,B}$ and $K_{e,W}$ are the extinction coefficients pertaining to bauxite and water, respectively. We assume that the bauxite and water can be separated into two distinct layers as in Fig. 36). Referring to Fig. 36, the heights $H_W$, and $H_B$ are given by

$$H_W = mH, \quad H_B = (1 - m)H, \tag{28}$$

where $m$ is the volume fraction of moisture in the bauxite. The values for $P(H)$, $P_0$, and $H$ are given in RAAL's data. In addition, we have several lab sample values for $m$. Substituting (28) into (27) and rearranging Eq. (27) leads to

$$-\frac{1}{H} \log\left(\frac{P(H)}{P_0}\right) = (K_{e,W} - K_{e,B})m + K_{e,B}\,. \tag{29}$$

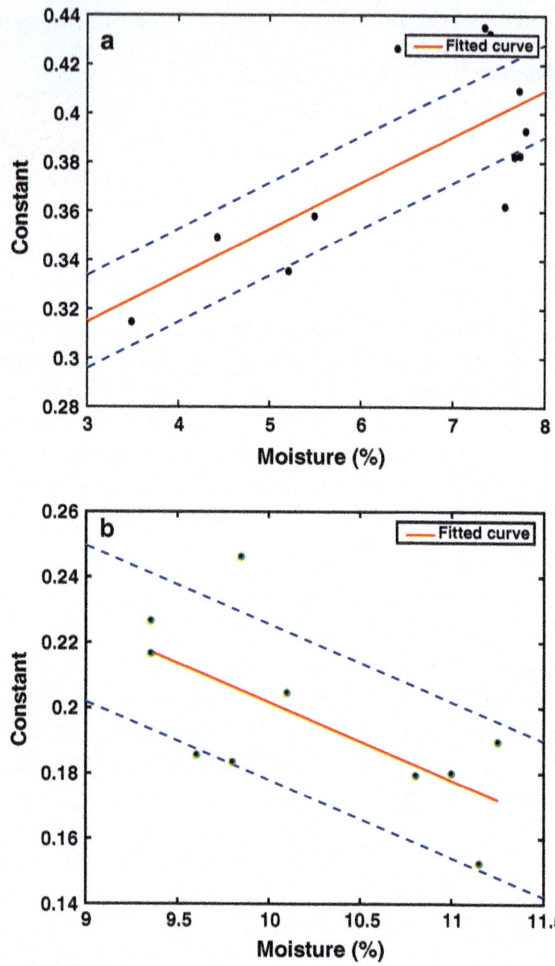


**Fig. 37** Left hand side of (29) against moisture content for **a** Hyena and **b** Maia ship data. The red lines correspond to the linear regression fits and the blue dashed lines are the related error bars. The black circles correspond to sample values from RAAL

This expression gives a direct relationship between the signal strength and the moisture content of the bauxite. We can think of (29) as being a linear regression equation where $(K_{e,W} - K_{e,B})$ is the slope of the line and $K_{e,B}$ is the y-intercept.

Figure 37 shows plots of the left hand side of (29) against moisture content. In particular, Fig. 37a, b pertain to data for the ships Hyena and Maia, respectively. The red lines are the linear regression fits and the blue dashed lines are the associated error bars. The black circles represent sample values from RAAL. We note that the Hyena and Maia results exhibit contradictory correlations between the LHS of (22) and moisture levels; the positive slope in Fig. 37a suggests that $K_{e,W} > K_{e,B}$ in the Hyena shipment, whilst the negative slope in Fig. 37b suggests that $K_{e,B} > K_{e,W}$ in the Maia shipment. These results indicate that either that different bauxites have significantly different effects on attenuation, or that the cancellation effects noted from interference with internal reflections cause attenuation to be an ineffective measure of moisture content in an environment where bauxite height varies with time. Whilst the linear fits do not give the best quantitative results, they do exhibit reasonably good qualitative agreement. This method does show some promise and with more work, perhaps can yield a good predictor for moisture content given bauxite height and the power levels of the analyser signal.

## 7 Discussion

In this report we summarise the various modelling approaches used during the Study Group. Our goal is to develop an understanding of the fundamental mechanisms occurring as the microwaves pass through the bauxite, and thus help to calibrate the analyser accordingly.

During the Study Group we adopted a two-fold approach, exploring the data for what it could tell us, and reviewing the differential equations and their solutions describing the theory for microwaves passing through complex polarising materials. We also tried some simpler approximate differential equation models.

The data was puzzling in a number of ways, most strikingly in the appearance of a squircle when the two-phase shifts were plotted one against the other. We now understand that the squircle is a consequence of the way that phase shift is measured, in the presence of noise.

We have showed that in the absence of noise, the squircle would sharpen at the corners to a parallelogram. We used this knowledge to obtain an algorithm to project the squircle onto the underlying parallelogram, thereby using one phase to reduce the noise in the other, and recovering more accurate phase shift measurements in the vicinity of the corners of the squircle, where one of the phases is near the critical values zero and $\pi$. We coded and showed this algorithm is successful using MATLAB.

The algorithm used to decide if $\phi$ is in a given line segment is given in the Appendix. The MATLAB code for the algorithms for lifting the phase shifts and for deciding if $\phi$ is in a given line segment has been placed with Zenodo and GitHub (McGuinness et al. 2019).

We observed that during time periods when the bauxite height was changing rapidly between zero and the typical value, the time resolution of one second was only just sufficient to resolve the changes in phase shift correctly through multiples of $\pi$ in value. The zero value, and how phases change as height changes from zero, are important for calibration purposes. We recommend either slower changes in bauxite depth or more rapid measurements (say four per second rather than one) to more carefully resolve these rapid signal changes.

We provided an algorithm, successfully coded in MATLAB, that uses the concurrent measurements of bauxite depth to lift the two multivalued measured phase shifts into one single-valued inferred true phase shift that can extend beyond $\pi$ and $2\pi$ in value. In fact, two lifted phase shifts can be obtained, and both look very similar as in Fig. 21. The output called squircle provided by the microwave analyser was seen (Fig. 23) to be very similar to our lifted phase, with the difference that our lifted phase is linear in actual phase whereas squircle has a sine wave feature. This is probably due to a computation in the analyser that uses measured phase angles to give an arclength from some reference position to a hand on a clock face which takes the squircle shape, and then calculating the angle of the clock hand, then outputting that angle as *squircle*.

The bilinear behaviour of phases during times of rapid change in bauxite depth was revealed to be due to the roughly half-second time offset of the depth measurement and the microwave measurement and was easily removed by shifting depth measurements by half a second (Fig. 28). This was done by hand but could be automated by choosing the lag to minimise variance in the phase versus height plots.

Attenuation data then looks cleaner as in Fig. 27b when plotted against properly lagged bauxite heights, and we found that the appearance of attenuation in such a plot is dominated by wave interference effects. Nevertheless, the attenuation of just the first signal does have the appearance in Fig. 30 of a straight line fitted to the combined signals, giving us an indication that despite the strong oscillatory effect of interference, if attenuation is plotted against height, a fitted straight line might be a reasonable proxy for the attenuation of a signal that does not experience interference. Furthermore, at heights between 200 and 300mm, the effect of interference is much reduced. This could explain the effectiveness of the results obtained using the Beer-Lambert law in Sect. 6.

Note that for the pelletised biomass case, since thickness does not vary, we do not expect that interference effects will be visible. They are a consequence of internal reflections at the top and bottom of the bauxite together with the distance travelled by the internally reflected waves, and for a constant thickness no such variation in interference will be observable.

The modelling of polarisers interacting with the microwaves provides strong theoretical links in Sect. 5 between moisture content and both phase shift and attenuation. This theoretical development indicates that both phase shift and attenuation are not in general linear in bauxite height, and hence on moisture content. The dependence on bauxite height is clearly linear in Fig. 35, but at a given height the spacing between lines bunches up near the high moisture side. For small changes in moisture content, a linear approximation will be effective enough.

For larger changes in moisture content, a nonlinear approach such as that demonstrated in Appendix B is seen to be very effective. In that appendix, phases are not projected, and the squircle is used directly to lift the phase. The bauxite used is drier than the Maia data used above, resulting in smaller overall phase shifts. The permittivity ratio with that of water is used directly **without calibration** to obtain a remarkably good measurement of moisture. With calibration, the measure improves a little. Using a fully nonlinear approach is seen to be effective for moisture levels varying from 4% to 8%.

It is remarkable that attenuation, according to the theory and as seen in the second plot in Fig. 35, reduces as moisture content increases, at a given height of bauxite. However, that theory ignores variations in conductivity $\sigma$ with water content, and only considers polarisability effects. This does not strongly affect phase shift, but the leading term in (25) is proportional to $\sigma/\sqrt{\epsilon_r}$.

Conductivity relates current density to the electric field intensity, and since the water in the bauxite is likely to contain a number of dissolved ions it may be similar to sea water with a relatively high bulk conductivity. However, the moisture is mostly distributed in a network of pores, with fluid paths that are not necessarily connected. There are usually critical values of porosity and liquid saturation above which percolation theory gives connected paths, with a power-law dependence of effective $\sigma$ on saturation near the critical values. We might expect in general that a higher moisture content gives a larger conductivity, which gives higher attenuation of the electric field.

Consequentially, we are mainly reliant on data to inform us of the overall relationship between moisture content and $\sigma/\sqrt{\epsilon_r}$ or attenuation. The data tells us two things—that interference effects from reflections from the bauxite/air and bauxite/belt interfaces are clearly visible when attenuation is plotted against bauxite height, and that attenuation does increase with moisture content as seen in Fig. 37, suggesting that conductivity effects dominate over permittivity effects.

Moisture levels in pelletised biomass have been successfully measured using microwaves (McKeown et al. 2017). In that application, the density of the biomass flow is a crucial variable, central to the analyses performed to infer moisture content. In the case of bauxite, density is not a key variable, although it does enter the theoretical formulae implicitly. Instead, density does not vary much compared to bauxite height, which is the critical variable for bauxite analogous to density for the pellets. Indeed, we find ourselves plotting phase change versus height, whereas in the pellets it is phase change versus flow density.

## 8 Conclusions

The key questions for the Study Group included the following. Is the analyser capable of measuring moisture accurately (within $+/-$ 1% moisture)? What is the best way to use the raw and processed data output by the analyser? Can the attenuation be used? Is density required? Is the measurement from the analyser bauxite-dependent?

We conclude that it is possible to use the analyser to measure moisture percentage within $+/-$ 1.5% at 95% confidence levels, provided care is taken to calibrate against lab samples and provided moisture content is not too high. A nonlinear approach would reduce the need for more frequent calibration.

The analysis that relates moisture content to phase shift here assumes a relatively small and constant value for the electrical conductivity of the wet bauxite. When moisture reaches critical values corresponding to connected pathways for (salty) water in the pore spaces, the analysis here becomes invalid. Furthermore, since conductivity is expected to dramatically increase at critical moisture levels, we expect the signal-to-noise ratio to become problematic too.

Interference effects, unique to the bauxite situation with its varying heights, dominate the appearance of attenuation versus height and render it much less useful than in the case of biomass pellets or grains where the thickness of the sample is unchanging. Phase shift is unaffected by interference, provided there is sufficient signal to measure phase.

The density function which is so important in the pelletised biomass case and in the case of grains and seeds is essentially replaced by variations in bauxite height, so direct measurement of density is not expected to be important for the bauxite case.

Different dry bauxite types will have different effects on microwave phase shift and attenuation, and these are captured by the theory we present. Calibration with lab moisture values should take care of these differences.

**Acknowledgements** We wish to acknowledge the support of Science Foundation Ireland for both a conference grant (SFI Grant 17/CW/4698) and a workshop grant (SFI 12/IA/1683).

## Appendix A: Locating a Point on a Line Segment

The algorithm used in Sect. 3.6 to decide if the tip of a vector $\boldsymbol{v}$ lies on a line segment AB is as follows:

Project $\boldsymbol{v}$ onto AB as

$$\boldsymbol{v}_{\text{proj}} = (\boldsymbol{v} - \boldsymbol{a}) \cdot (\boldsymbol{b} - \boldsymbol{a}) \frac{(\boldsymbol{b} - \boldsymbol{a})}{||(\boldsymbol{b} - \boldsymbol{a})||^2}$$

where $\boldsymbol{a}$ is the vector from origin to the point A, and $\boldsymbol{b}$ is the vector from origin to the point B, and $||\boldsymbol{v}||$ is the norm or length of the vector $\boldsymbol{v}$. Then the vector from the tip of the vector $\boldsymbol{v}$ to the line through the points A and B, and orthogonal to that line, is given by

$$\boldsymbol{v}_{\text{orth}} = \boldsymbol{v} - \boldsymbol{a} - \boldsymbol{v}_{\text{proj}} \ .$$

If the length or norm of $\boldsymbol{v}_{\text{orth}}$ is greater than some small tolerance value, we conclude that $\boldsymbol{v}$ is not on the line segment AB.

Otherwise, $\boldsymbol{v}$ is in the line that goes through both A and B, but might not be between A and B. So we find the parameter $t$ in the vector form of the projection vector

$$\boldsymbol{v}_{\text{proj}} = \boldsymbol{a} + t(\boldsymbol{b} - \boldsymbol{a})$$

and noting that the line segment AB corresponds to $t \in [0, 1]$, we check that $t$ lies in the range 0, 1. If so, we conclude that the tip of vector $\boldsymbol{v}$ lies on a line segment AB, within some small tolerance.

The MATLAB code for the algorithms for lifting the phase shifts and for deciding if $\phi$ is in a given line segment, has been placed with Zenodo and GitHub (McGuinness et al. 2019).

## Appendix B: A Simpler Approach—No Projection But Nonlinear

In this section the treatment of phase shifts is fully nonlinear. Thus the refractive index and relative permeability are treated as real quantities. The two instrumental phase shifts are not projected onto a parallelogram before lifting them to find a true phase shift value, they are used raw. Good results are seen to be obtained without any projection of phase shifts. The data used in this section is from the ship Heina. It is notably drier and covers a larger moisture range than the Maia case studied above. The total number of data points is approximately $4 \times 10^5$, collected over about 100 h of bauxite loading. Notably, a good match to lab values of moisture content is obtained using simple approximations without any calibration to lab values.

Note that the correspondence here between moisture content and phase shift is nonlinear in phase shift, since permittivity is calculated nonlinearly rather than regressing moisture against phase shift directly.

The basic physics the device is based on is shown in Fig. 38. As microwaves with wavenumber $k$ pass through a thickness $L$ of a medium with a refractive index $n$ they pick up an additional phase $\Delta\phi = (n-1)kL$ compared to microwaves that pass through a medium of refractive index $n = 1$, e.g. air or vacuum.

The following steps are carried out to extract an estimate of water content from the phase shift data:

- The raw $\phi_1, \phi_2$ data is linearly scaled to approximately fit within the range $[0 : 1000] \times [0 : 1000]$.
- A phase $\phi$ is extracted from the transformed $\phi_1, \phi_2$.
- The refractive index and thus the dielectric constant are extracted from the gradient of a graph of $\phi$ as a function of $h$.
- Moisture content is estimated from the dielectric constant.

Phase information is encoded as $\phi_1, \phi_2$ coordinates. The unprocessed $\phi_1, \phi_2$ data is shown in Fig. 39. These data are processed by a linear transformation to a squircle with equal side lengths as shown in Fig. 40. The linear transformation is of the form $\boldsymbol{\phi} \to A\boldsymbol{\phi} + B$, where $\boldsymbol{\phi} = (\phi_1, \phi_2)$, and where $A$ is a real matrix and $B$ is a real vector.

In particular, for Fig. 40, the following calculations were made:

$$\boldsymbol{\phi}_A = \begin{pmatrix} \frac{1}{600} & \frac{1}{600} \\ \frac{1}{400} & -\frac{1}{400} \end{pmatrix} \boldsymbol{\phi} + \begin{pmatrix} \frac{-945}{600} \\ \frac{20}{400} \end{pmatrix}$$

then rescale $\boldsymbol{\phi}_A$ to obtain the processed angle

$$\boldsymbol{\phi}_{\texttt{processed}} = 250 \begin{pmatrix} 1 & -1 \\ 1 & 1 \end{pmatrix} \boldsymbol{\phi}_A + 500$$

The phase $\phi$ is extracted from the processed values of $\phi_1$ and $\phi_2$ by assuming that $\phi$ is equal to distance travelled around the squircle and one complete loop of the

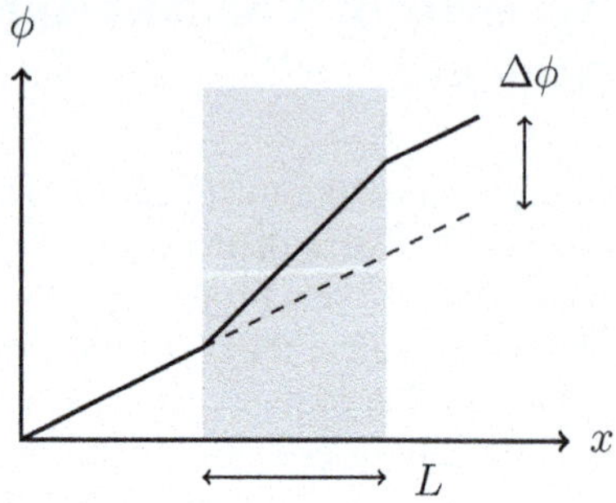


**Fig. 38** As the microwave radiation passes through the bauxite it picks up an extra phase $\Delta\phi = (n-1)kL$ relative to a reference signal that does not pass through bauxite

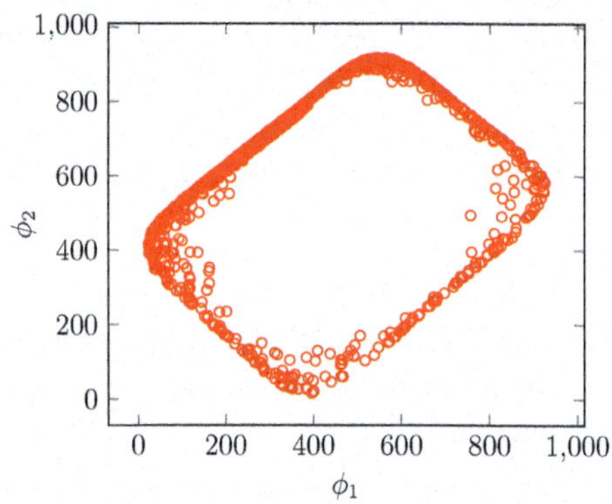


**Fig. 39** Unprocessed $\phi_1, \phi_2$

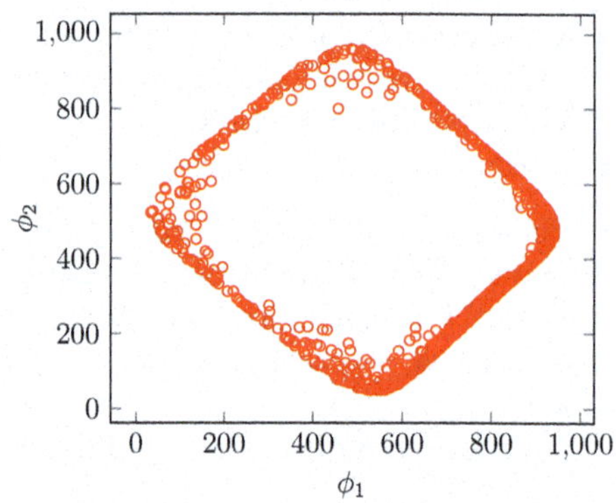


**Fig. 40** Processed $\phi_1, \phi_2$

squircle corresponds to $2\pi$. This is shown in Figs. 41, 42 and 43. The algorithm used to compute the phase using $\phi_1$ is (dropping the subscript `processed`)

$$\phi = \begin{cases} \frac{\pi\phi_1}{1000} & , \quad \phi_2 > 500 \\ 2\pi - \frac{\pi\phi_1}{1000} & , \quad \phi_2 < 500 \,. \end{cases}$$

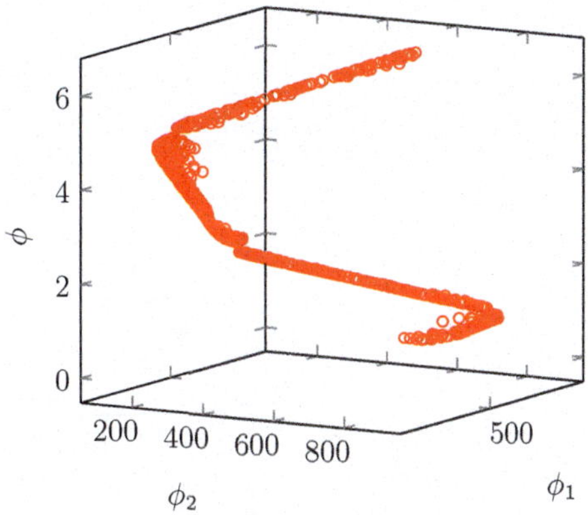


**Fig. 41** Phase $\phi$ extracted from $\phi_1$ and $\phi_2$

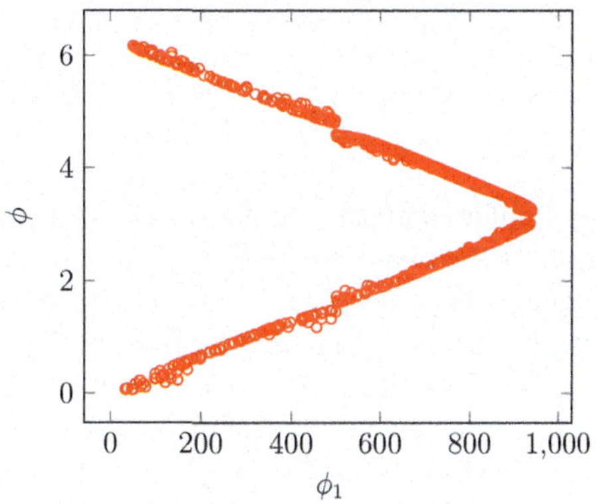


**Fig. 42** Phase $\phi$ plotted against $\phi_1$

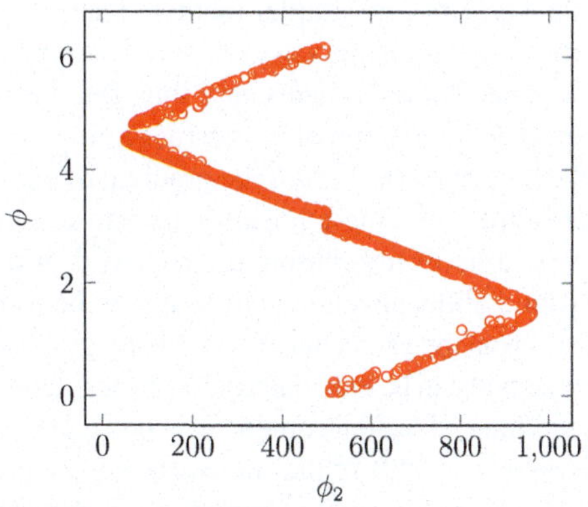


**Fig. 43** Phase $\phi$ plotted against $\phi_2$

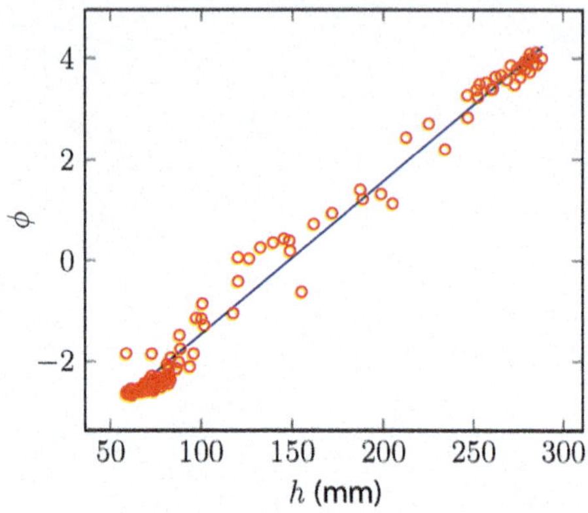


**Fig. 44** A graph of $\phi$ against bauxite height $h$ together with a best fit line, for Heina data chosen over an arbitrary time interval but one with a full range of bauxite heights $h$ attained. Note that the $\phi$ data has been lifted or unwrapped, to extend over a range of nearly $3\pi$, leveraging bauxite height, in a similar manner to the development in Sect. 3.5 but without projecting

A similar approach was also used, switching $\phi_1$ and $\phi_2$, and it produced very similar values for phase change. It is clear from the algorithm that *distance* around the squircle here is a change in $\phi_i$, not an arclength. A change in one of the $\phi_i$ values is what is needed to prove a measure of the phase shift.

This provides a lifted $\phi$ as does the method that uses projection and the height of the bauxite, but the present method is limited to phase changes that are no larger than $2\pi$. The limitation is because the information provided by bauxite height is not used here, only the (processed) values of $\phi_1$ and $\phi_2$. Using projection is slightly better than the present, simpler method, because projection removes the bias from variance due to the XOR process.

The refractive index and thus the dielectric constant can be extracted from the gradient of a graph of $h$ vs $\phi$ (assuming the moisture content is only slowly varying). Since $\Delta\phi = (n-1)kh$ the gradient of a graph of $\phi$ as a function of $h$ is $(n-1)k$. An example showing a linear fit to data is shown in Fig. 44. (An alternative procedure would be to fit a baseline curve to $\phi$ at times when the conveyor was empty: this would allow absolute values of $\phi$ to be used.)

From the gradients of $\phi$ against $h$ graphs the refractive index and thus the dielectric constant can be calculated ($\epsilon = n^2$ if $\epsilon, n \in \mathcal{R}$), this is shown in Fig. 45. A very simple estimate of water content can be made by linear interpolation: $(\epsilon - 1)/(\epsilon_w - 1)$, where $\epsilon_w = 80$ is the dielectric constant of pure water. As Fig. 46 shows, this is already a reasonable estimate of water content. An improved version created by a linear calibration to the data is shown in Fig. 47.

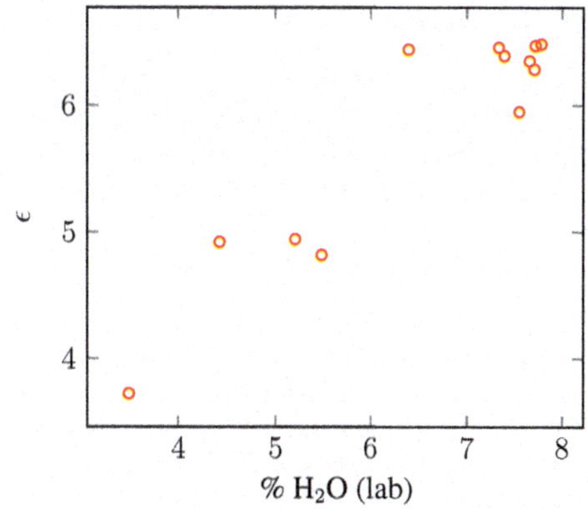


**Fig. 45** Graph of relative dielectric constant $\epsilon$ versus measured water content

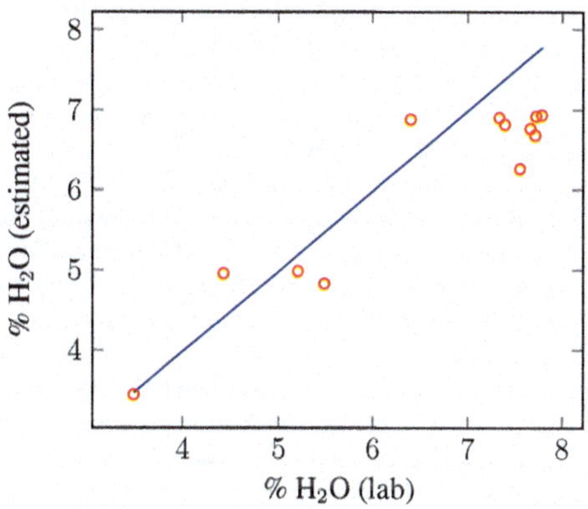


**Fig. 46** Water content estimated as $\epsilon/\epsilon_w$ versus measured water content. The blue line shows a perfect result

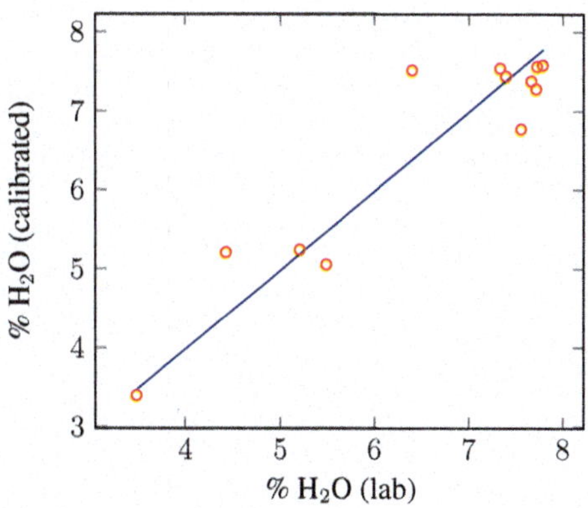


**Fig. 47** Calibrated water content versus measured water content. The blue line shows a perfect result. Lab measurements have been lagged by twelve minutes to ensure the sampling times correspond to analyser times

## References

Kraszewski A, Kulinski S (1976) An improved microwave method of moisture content measurement and control. IEEE Trans Ind Electron Control Instrum IECI-23(4):364–370 (1976)

Landau LD, Lifshitz EM (1984) Electrodynamics of continuous media, vol. 8 of course of theoretical physics, 2nd edn. Pergamon Press. Translated from Russian

Lorrain P, Corson DR, Lorrain F (1987) Electromagnetic fields and waves, 3rd edn. Freeman and Co., NY

McGuinness M, Bohun S, Cregan V, Lee WT, O'Keeffe G, Dewynne J (2019) Functions used in continuous detection of ore moisture using microwaves (January 2019) [cited Jan 2019]. https://zenodo.org/record/2532261

McKeown MS, Trabelsi S, Nelson SO, Tollner EW (2017) Microwave sensing of moisture in flowing biomass pellets. Biosyst Eng 155:152–160

Nelson SO (1987) Models for the dielectric constants of cereal grains and soybeans. J Microwave Power Electromag Energy 22(1):35–39

Nelson SO (1992) Correlating dielectric properties of solids and particulate samples through mixture relationships Trans. ASAE 35:625–9

Nelson SO, Trabelsi S (2010) Measurement of grain and seed microwave permittivity for moisture and density determination. In: Proceedings of the IEEE SoutheastCon 2010 (SoutheastCon). IEEE, pp 463–466. Web

Trabelsi S, Kraszewski AW, Nelson SO (1997) A new density-independent function for microwave moisture content determination in particulate materials. In: IEEE instrumentation measurement technical conference proceedings. Ottawa, Canada, Mary, vol 1, pp 648–652

Trabelsi S, Kraszewski AW, Nelson SO (1998) A microwave method for online determination of bulk density and moisture content of particulate materials. IEEE Trans Instrum Meas 47(1):127–132

Trabelsi S, Kraszewski AW, Nelson SO (2000) Phase-shift ambiguity in microwave dielectric properties measurements. IEEE Trans Instrum Meas 49(1):56–60

Trabelsi S, Kraszewski AW, Nelson SO (2001) New calibration technique for microwave moisture sensors. IEEE Trans Instrum Meas 50(4):877–881

Viana SAA (2013) Microwave-based moisture measurement of bauxite ore on a conveyor belt. Eng Mining J. Features

# The Magnetic Field About a Three-Dimensional Cylindrical Magnet

**Graham Weir, Jerome Leveneur, and Nicholas Long**

**Mathematics Subject Classification** MCS2000 primary 74C99 and secondary 52C35 and 74A45

## 1 Introduction

Neodymium (Kaye et al. 1995) magnets (also called NdFeB, NIB or Neo magnets) are the most widely used rare-earth magnet because their strong magnetic field favours industrial applications where weight and size are important factors. They are often sold as rectangular prisms or cylinders, and sometimes as spheres, or as other shapes. The theory of the magnetic field from a rectangular neodymium magnet has been published (Vacquier et al. 1951; Weir et al. 2020). This paper considers the magnetic field from cylindrical magnets.

We assume that the cylindrical magnet is uniformly magnetised along its axis because during manufacture, a large magnetic field is applied normally to sheets of neodymium to magnetise them. Then the magnets are formed from these magnetised sheets. The assumption of uniform magnetisation is often made in mathematical models (Corson and Lorrain 1962).

Magnets will never have a completely uniform magnetisation because of variations in the size, orientation, strength and chemical composition of the magnetic domains. The typical size of a magnetic domain is around $10^{-5}$ m, which is small relative to the typical size of a neodymium magnet, of around $10^{-2}$ m. We expect then that much

G. Weir (✉)
Massey University Manawatu, Palmerston North 4442, Palmerston North, New Zealand
e-mail: grahamweir@xtra.co.nz

J. Leveneur
GNS Science, Lower Hutt 5040, Lower Hutt, New Zealand
e-mail: j.leveneur@gns.cri.nz

N. Long
Victoria University of Wellington, Petone 5046, Wellington, New Zealand
e-mail: nick.long@vuw.ac.nz

 
R. McKibbin et al. (eds.), *Proceedings of the Forum "Math-for-Industry" 2019*, Mathematics for Industry 36, https://doi.org/10.1007/978-981-19-1154-5_3

of the variability resulting from domain structure will average out, and the magnetic field can be approximated as resulting from a uniformly magnetised body.

If an application uses several cylindrical magnets, then the results from this paper will be able to model the resulting field because the corresponding Maxwell equations are linear. We show below that the components of the magnetic field are described by a finite number of complete elliptic integrals, of the first, second and third kinds.

The magnetic field from a finite magnetic body rapidly approaches that of a point dipole field, for field points far from the magnet. Because many industrial uses of neodymium magnets will require knowledge of the magnetic field close to the magnetic body, the exact expressions presented here should be relevant, especially because of the widespread use of cylindrical neodymium magnets in industrial applications.

Identifying accurate analytical solutions to the behaviour of magnetic objects and systems has useful industrial and scientific applications in a wide range of fields. Understanding the magnetic behaviour at the surface of magnetic materials is particularly important to analyse and interpret data from Magnetic Force Microscopy (MFM) (Folks and Woodward 1998; Nevcas et al. 2019) and magneto-optic Kerr effect microscopy (Won et al. 2005; Göddenhenrich et al. 1988). Using these microscopy techniques, field reversal is commonly observed about the surface and often ascribed to magnetic domain structure and pinning site (Hartmann 1999; Schwarz and Wiesendanger 2008). In other applications where an accurate knowledge of magnet behaviour is required, such as microfluidics (Cao et al. 2014; Grant et al. 2002), mineral exploration (Shaar and Feinberg 2013; Sattel and Reid 2006), particle accelerator (Yuan et al. 1987; Melone et al. 2012) and plasma confinement systems (Helander et al. 2020; Goncharov and Brown 2007), analytical solutions of the magnetic field about magnetic materials and their magnetisation can assist design of better performing systems. Finally, the exact results in this paper may be useful for validating computer code, which calculates two- and three-dimensional magnetic fields.

## 2 The 3D Magnetic Field

In a Cartesian coordinate system, the steady magnetic field **B** from a cylindrical magnet with uniform magnetisation **M** along the cylinder axis is Weir et al. (2020)

$$\mathbf{B}(\mathbf{x}) = \frac{\mu_0}{4\pi} \int dS' \frac{((\mathbf{M} \times \mathbf{n}) \times \mathbf{e}_r)(\mathbf{x}')}{|\mathbf{x} - \mathbf{x}'|^2} \tag{1}$$

where $\mathbf{n}$ is the unit normal to the magnet surface, $\mathbf{e}_r$ is the unit vector from an element of the surface of the magnet, to the field or observation point, $\mathbf{x}$ is the location of a field point centred on the centre of the magnet, $\mathbf{x}'$ is a source point on the surface of the magnet, the surface integral over $dS'$ covers all of the magnet surface and

$$(\mathbf{M} \times \mathbf{n}) \times \mathbf{e}_r = \mathbf{n}(\mathbf{M} \cdot \mathbf{e}_r) - \mathbf{M}(\mathbf{n} \cdot \mathbf{e}_r) \tag{2}$$

where the notation used in this paper is given in Table 1.

**Table 1** Notation used in this paper

| Symbol | Meaning | Units |
|---|---|---|
| $a$ | Cylinder radius | m |
| $\mathbf{b}$ | Non-dimensional magnetic induction from a disk | – |
| $\mathbf{B}$ | Magnetic induction | Wb m$^{-2}$ |
| $B_R$ | $\mathbf{B} \cdot \mathbf{e}_R$ | Wb m$^{-2}$ |
| $\mathbf{e}_n$ | Unit vector normal to cylinder surface | – |
| $\mathbf{e}_r$ | Unit vector from a source to field location | – |
| $\mathbf{e}_R$ | Unit vector in cylindrical $R$ direction | – |
| $\mathbf{e}_x$ | Unit vector in x direction | – |
| $\mathbf{e}_y$ | Unit vector in y direction | – |
| $\mathbf{e}_z$ | Unit vector in $\mathbf{M}$ direction | – |
| $\mathbf{e}_\phi$ | $\mathbf{e}_z \times \mathbf{e}_n$ | – |
| $E$ | Complete elliptic of the second kind | – |
| $H$ | Cylinder height | m |
| $k_-$ | Elliptic modulus | – |
| $k_+$ | Elliptic modulus | – |
| $K$ | Complete elliptic of the first kind | – |
| $\mathbf{m}$ | Magnetic dipole moment | A m$^2$ |
| $M$ | $\lvert\mathbf{M}\rvert$ | A-turns m$^{-1}$ |
| $\mathbf{M}$ | Magnetisation vector | A-turns m$^{-1}$ |
| $n$ | Elliptic characteristic | – |
| $\mathbf{n}$ | Outward unit normal to magnet surface | – |
| $r$ | $\lvert\mathbf{x} - \mathbf{x}'\rvert$ | m |
| $r_+$ | Maximum distance to lower magnet surface | m |
| $r_-$ | Maximum distance to upper magnet surface | m |
| $R$ | $\sqrt{x^2 + y^2}$ | m |
| $\mathbf{x}$ | 3D field location, centred on cylinder centre | m |
| $\mathbf{x}_m$ | midpoint of magnet | m |
| $\mathbf{x}'$ | 3D source location | m |
| $z$ | Vertical coordinate along $\mathbf{M}$ direction | – |
| $\epsilon$ | Small non-dimensional variable | – |
| $\mu_0$ | Permeability of vacuum ($4\pi \times 10^{-7}$) | H m$^{-1}$ |
| $\chi$ | $\phi - \psi$, angular variable | – |
| $\phi$ | Cylindrical angular variable, $\tan\phi = y'/x'$ | – |
| $\psi$ | Cylindrical angular variable, $\tan\psi = y/x$ | – |
| $\Pi$ | Complete elliptic of the third kind | – |

A physical interpretation of (1) is that this magnetic field is generated by an apparent surface current $\mathbf{M} \times \mathbf{n}$, flowing azimuthally around the surface of the magnet. If $\mathbf{M}$ varies spatially, then an additional volume integral involving $\nabla \times \mathbf{M}$ results, and the magnetic field will not originate wholly from the surface contribution of an equivalent current.

The unit vector **n** can be resolved into cartesian coordinates, using

$$\mathbf{n} = (\cos\phi)\mathbf{e}_x + (\sin\phi)\mathbf{e}_y; \tag{3}$$

where $\phi$ is the angular variable in a cylindrical coordinate system, aligned with the cylinder geometry. The magnetisation vector can be written as

$$\mathbf{M} = M\mathbf{e}_z \tag{4}$$

while the unit vector $\mathbf{e}_r$ is

$$\mathbf{e}_r = \left(\frac{x - x'}{r}, \frac{y - y'}{r}, \frac{z - z'}{r}\right); \quad r = |\mathbf{x} - \mathbf{x}'|. \tag{5}$$

From (1)–(5), the vertical component of **B**, $B_z$, is

$$B_z(\mathbf{x}) = -\frac{\mu_0 M a}{4\pi}\int_0^{2\pi} d\phi[(x - x')\cos\phi + (y - y')\sin\phi]\int_{-H/2}^{H/2}\frac{dz'}{r^3}. \tag{6}$$

Assuming temporarily that the field point is not on the magnet surface, then (Gradshteyn and Ryzhik 1980)

$$\int\frac{dz'}{r^3} = \frac{-(z - z')}{r[(x - x')^2 + (y - y')^2]}; \quad x' = a\cos\phi; \quad y' = a\sin\phi \tag{7}$$

and so the angular integral in (6) involves the grouping $(x\cos\phi + y\sin\phi) = R\cos(\phi - \psi)$, where $R^2 = x^2 + y^2$, and $\tan\psi = y/x$. Writing $d\phi = d(\phi - \psi)$ allows $\psi$ to be eliminated from the angular integral, yielding

$$B_z(\mathbf{x}) = \frac{\mu_0 M}{4\pi}\int_0^{2\pi}\frac{a(z - \frac{H}{2})(R\cos\phi - a)d\phi}{(R^2 + a^2 - 2aR\cos\phi)\sqrt{R^2 + a^2 + (z - \frac{H}{2})^2 - 2aR\cos\phi}} \tag{8}$$

$$-\frac{\mu_0 M}{4\pi}\int_0^{2\pi}\frac{a(z + \frac{H}{2})(R\cos\phi - a)d\phi}{(R^2 + a^2 - 2aR\cos\phi)\sqrt{R^2 + a^2 + (z + \frac{H}{2})^2 - 2aR\cos\phi}}. \tag{9}$$

From (9), on the segment $[0, \pi]$ define $\phi = \pi - 2\theta$, while on the segment $[\pi, 2\pi]$ define $\phi = \pi + 2\theta$, and obtain

$$\int_0^{2\pi} f(a - b\cos\phi)d\phi = 4\int_0^{\frac{\pi}{2}} f(a + b\cos 2\theta)d\theta = 4\int_0^{\frac{\pi}{2}} f(a + b - 2b\sin^2\theta)d\theta. \tag{10}$$

From (9)–(10),

$$B_z(\mathbf{x}) = \frac{\mu_0 M(z - \frac{H}{2})}{2\pi r_-}\left[\frac{R-a}{R+a}\Pi(n|k_-) - K(k_-)\right] \tag{11}$$

$$- \frac{\mu_0 M(z + \frac{H}{2})}{2\pi r_+}\left[\frac{R-a}{R+a}\Pi(n|k_+) - K(k_+)\right] \tag{12}$$

$$r_\pm = \sqrt{(R+a)^2 + \left(z \pm \frac{H}{2}\right)^2};\quad r_0 = \sqrt{(R+a)^2 + \left(\frac{H}{2}\right)^2} \tag{13}$$

where $r_+$ (respectively, $r_-$) is the distance from the field point to the furthest point on the lower (respectively, upper) horizontal magnet surface.

The complete elliptic integrals of the first, second and third kind, $E(k)$, $K(k)$ and $\Pi(n|k)$ respectively, are (Whittaker and Watson 1927; Byrd and Friedman 1954)

$$K(k) = \int_0^{\frac{\pi}{2}} \frac{d\theta}{\sqrt{1 - k^2\sin^2\theta}};\quad E(k) = \int_0^{\frac{\pi}{2}} d\theta\sqrt{1 - k^2\sin^2\theta} \tag{14}$$

$$\Pi(n|k) = \int_0^{\frac{\pi}{2}} \frac{d\theta}{(1 - n\sin^2\theta)\sqrt{1 - k^2\sin^2\theta}}. \tag{15}$$

Here $n$ is the elliptic characteristic, $k$ is the elliptic modulus, $k^2$ is the elliptic parameter, and $\alpha(\sin\alpha = k)$ the modular angle. The cylindrical magnet has one elliptic characteristic $n$, and two elliptic parameters $k_\pm^2$:

$$n = \frac{4aR}{(R+a)^2};\quad k_\pm^2 = \frac{4aR}{(R+a)^2 + \left(z \pm \frac{H}{2}\right)^2};\quad k_0^2 = \frac{4aR}{(R+a)^2 + \left(\frac{H}{2}\right)^2} \tag{16}$$

Note that $k_+(k_-)$ relates to the lower (upper) horizontal magnet surface.

The complete elliptic integrals in this paper were evaluated using Python software. The scipy.special library contains ellipe and ellipk, which evaluate the complete elliptic integrals of the first and second kind, respectively. The mpmath.ellippi package evaluates the complete elliptic integral of the third kind.

The elliptic integrals in (12) are divergent for $n = 1$ and for $k_\pm = 1$. When $R = a$, $n = 1$, while $k_\pm = 1$ when $R = a$ and $z = \mp H/2$; i.e. when the field position lies on the upper or lower magnet edge. However, the terms $(z \pm \frac{H}{2})$ in (12) then cancel the corresponding singularities, ensuring the vertical component $B_z$ is bounded everywhere.

The analysis for the radial component of **B** is similar to that for $B_z$. Because

$$\int \frac{(z - z')dz'}{r^3} = \frac{1}{r} \tag{17}$$

and using (1)–(5), the component $B_x$ is

$$B_x(\mathbf{x}) = \frac{\mu_0 Ma}{4\pi} \int_0^{2\pi} \left[ \frac{d\phi \cos\phi}{\sqrt{R^2 + a^2 + \left(z - \frac{H}{2}\right)^2 - 2a(x\cos\phi + y\sin\phi)}} \right. \tag{18}$$

$$\left. - \frac{d\phi \cos\phi}{\sqrt{R^2 + a^2 + \left(z + \frac{H}{2}\right)^2 - 2a(x\cos\phi + y\sin\phi)}} \right]. \tag{19}$$

Similarly, the component $B_y$ is

$$B_y(\mathbf{x}) = \frac{\mu_0 Ma}{4\pi} \int_0^{2\pi} \left[ \frac{d\phi \sin\phi}{\sqrt{R^2 + a^2 + \left(z - \frac{H}{2}\right)^2 - 2a(x\cos\phi + y\sin\phi)}} \right. \tag{20}$$

$$\left. - \frac{d\phi \sin\phi}{\sqrt{R^2 + a^2 + \left(z + \frac{H}{2}\right)^2 - 2a(x\cos\phi + y\sin\phi)}} \right]. \tag{21}$$

From symmetry, the current source is wholly azimuthal, and so the azimuthal component of **B** is zero. The non-zero horizontal component of **B** is $B_R = B_x \cos\psi + B_y \sin\psi$. Defining $\chi = \phi - \psi$ gives

$$B_R(\mathbf{x}) = \frac{\mu_0 Ma}{4\pi} \int_0^{2\pi} \left[ \frac{d\chi \cos\chi}{\sqrt{R^2 + a^2 + \left(z - \frac{H}{2}\right)^2 - 2aR\cos\chi}} \right. \tag{22}$$

$$\left. - \frac{d\chi \cos\chi}{\sqrt{R^2 + a^2 + \left(z + \frac{H}{2}\right)^2 - 2aR\cos\chi}} \right] \tag{23}$$

which can be expressed as complete elliptic integrals of the first and second kind,

$$B_R(\mathbf{x}) = \frac{\mu_0 M r_-}{2\pi R}\left[\frac{R^2 + a^2 + \left(z - \frac{H}{2}\right)^2}{(R+a)^2 + \left(z - \frac{H}{2}\right)^2} K(k_-) - E(k_-)\right] \tag{24}$$

$$- \frac{\mu_0 M r_+}{2\pi R}\left[\frac{R^2 + a^2 + \left(z + \frac{H}{2}\right)^2}{(R+a)^2 + \left(z + \frac{H}{2}\right)^2} K(k_+) - E(k_+)\right]. \tag{25}$$

The elliptic integrals $K(k_\pm)$ in (25) are logarithmically divergent for $k_\pm = 1$, or when $R = a$ and $z = \mp H/2$. Hence the radial component $B_R$ is unbounded when the field position lies on the upper or lower magnet edge. Therefore, the highest (lowest) point on the magnetic field line through the upper (lower) edge occurs at the upper (lower) edge of the magnet.

## 3 Properties of the Magnetic Field

The aim of this section is to describe some of the properties of the magnetic field obtained in Sect. 2.

**The axis R = 0**. From symmetry, the magnetic field is wholly vertical along the axis of the magnet $R = 0$, and (9) gives

$$B_z(0, z) = \frac{\mu_0 M}{2}\left[\frac{(z + \frac{H}{2})}{\sqrt{a^2 + \left(z + \frac{H}{2}\right)^2}} - \frac{(z - \frac{H}{2})}{\sqrt{a^2 + \left(z - \frac{H}{2}\right)^2}}\right]. \tag{26}$$

The maximum of $B_z(R, z)$ on the axis occurs at the magnet midpoint, where

$$B_z(0, 0) = \frac{\mu_0 M H}{\sqrt{H^2 + 4a^2}} \tag{27}$$

while $B_z$ on the upper or lower surface, on the axis, equals

$$B_z\left(0, \pm\frac{H}{2}\right) = \frac{\mu_0 M H}{2\sqrt{H^2 + a^2}}. \tag{28}$$

The magnetic field formed by removing or replacing cylindrical sections of magnetised material by non-magnetised material, from a cylindrical magnet, follows from superposition of the results above. The axial magnetic field from a coaxial cylinder, $0 < a_0 \leq R \leq a$, formed by removing a central cylindrical section of radius $a_0$ from a cylindrical magnet, has its field lines looping down through the magnet, since from (27), the magnetic field then is everywhere negative along the axis $z = 0$.

This example may help to clarify the general concept, originating from Ampere, that a fixed magnetic field is equivalent to the field from a collection of current loops.

This is clearly untrue for a solid cylindrical magnet, which has its central field line pointing upwards, whereas the corresponding current loop has its central magnetic field line pointing downwards, in the symmetry plane of the loop.

The previous two paragraphs show that the magnetic field from a coaxial cylinder, even with an arbitrarily small internal radius, does indeed look like that from a current loop. However, once the removed magnetic material is replaced, the magnetic field changes dramatically, emphasising that Ampere's concept is true, subject to the shape of the magnetic material.

If, instead of removing or replacing all of the magnetised material along the axis of a cylindrical magnet, only a very small cylindrical section of material is removed or replaced from the axis below the upper surface, then from the paragraph of (39), the effects of this change will be confined to a length scale determined by the cube root of the volume of material removed. However, the effect on the field about the upper surface can be drastic, perhaps reversing the sign of the magnetic field there.

From (27), the maximum value of $B_z$ at the centre of the small inner cylinder is greater (smaller) than that for the larger cylinder if $a_0/h < a/H$ ($a_0/h > a/H$). Consequently removing magnetised material has an enhanced effect on the local magnetic field, if its ratio of height to width is greater than that of the larger cylinder.

If a change in magnetisation of $\Delta M$ occurs over a small cylinder of radius $a_0$ and height $h$, aligned along the magnet axis, and whose upper surface coincides with that of the magnet, then the change $\Delta B_z$ in the axial magnetic field is

$$\Delta B_z(0,z) = \frac{\mu_0 \Delta M}{2}\left[\frac{(z+h-\frac{H}{2})}{\sqrt{a_0^2+\left(z+h-\frac{H}{2}\right)^2}} - \frac{(z-\frac{H}{2})}{\sqrt{a_0^2+\left(z-\frac{H}{2}\right)^2}}\right]. \tag{29}$$

This is plotted in Fig. 1, showing that local increases (decreases) in magnetisation of 40 and 25% can result in doubling (reversing) the local axial magnetic field, for moderate shape ratios $h/a_0$, which are greater than that for the magnet, $a/H$.

**The plane z = 0**. From (25), $B_R$ is zero on the plane $z = 0$. $B_z$ is positive close to the axis but becomes negative far from the axis. From (12), $B_z$ is zero on the plane $z = 0$, corresponding to the fixed points of **B**, when

$$\frac{(R-a)}{(R+a)}\Pi(n|k_0) = K(k_0); \quad \frac{1}{n} = \frac{1}{k_0^2} - \frac{H^2}{16aR} \tag{30}$$

where $k_0$ is defined in (16). The "solution" to (30) is $R = a$, where $\Pi(n|k_0)$ is singular because there $n = 1$.

This result that the fixed points of **B** lie in the magnet surface can be understood because from (1), the magnetic field results from the surface current $\mathbf{M} \times \mathbf{n}$. The fixed points of the corresponding magnetic field will then lie inside this surface current, and hence in the magnet surface.

Alternatively, from (30), the limit as $n$ tends to unity, while $k_0$ is fixed away from unity gives

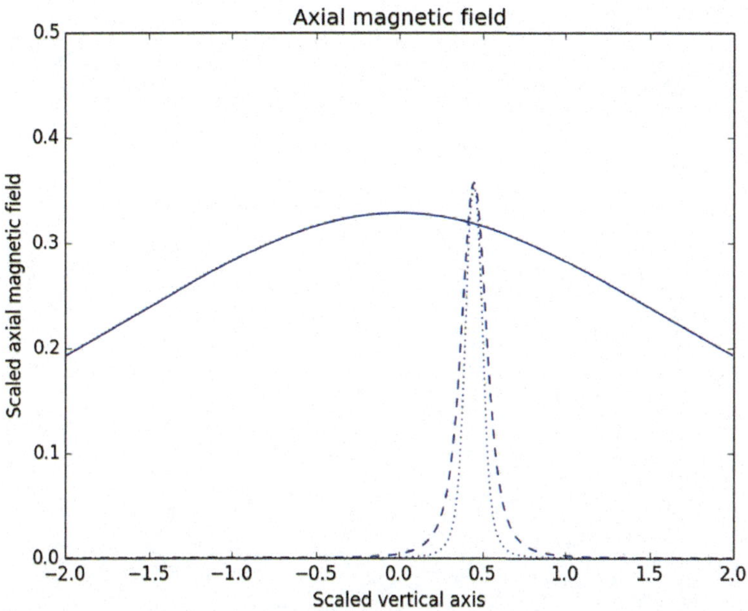


**Fig. 1** Scaled axial magnetic field from (26), for $a = 3$, $H = 1$ (solid curve); and from (29) for $\Delta M = 0.4M, a_0 = 0.1, h = 0.1$ (broken curve); $\Delta M = 0.25M, a_0 = 0.05, h = 0.1$ (dotted curve), when all upper surfaces are at $z = H/2$

$$\Pi(n|k_0) \simeq \frac{\pi\sqrt{n}}{2\sqrt{(n-m)(1-n)}}; \quad \frac{|(R-a)|}{(R+a)}\Pi(n|k_0) \simeq \frac{2\pi a}{H} \tag{31}$$

and so for $R$ close to $a$, on $z = 0$,

$$B_z(a^{\pm}, 0) \simeq \mu_0 M \left[\frac{HK(k_0)}{4\pi a} \mp \frac{1}{2}\right]; \quad K(k_0) \simeq \ln\left(\frac{16a}{H}\right). \tag{32}$$

From (32), $B_z$ undergoes a negative jump of $\mu_0 M$ on passing outwards across the magnet boundary on $z = 0$ because the azimuthal surface current density equals $M$ there. For typical neodymium magnets, (32) shows that $B_z$ is positive (negative) on $z = 0$, for $R < a$ ($R > a$). The concept of the fixed points of the magnetic field then is ambiguous because the magnetic field is discontinuous at $(R, z) = (a, 0)$, and the mean value of $B_z$ across the jump, is positive.

From (16), $k_0^2$ differs from $n$ by the term $(H/2)^2$, which is usually relatively small. Substituting $k_0^2 = n$ into (30) yields unbounded values at $R = a$. Consequently, although the terms involving $(H/2)^2$ in (30) are typically negligible, they become important around the boundary of the magnet, on $z = 0$.

Figure 2 plots the magnetic field lines, about the edge of a magnet with $a = 1.$ and $H = 0.25$. The field lines are horizontal at $(R, z) = (a, H/2)$ and increase rapidly in slope on moving across the upper magnet surface, away from the edge of the magnet.

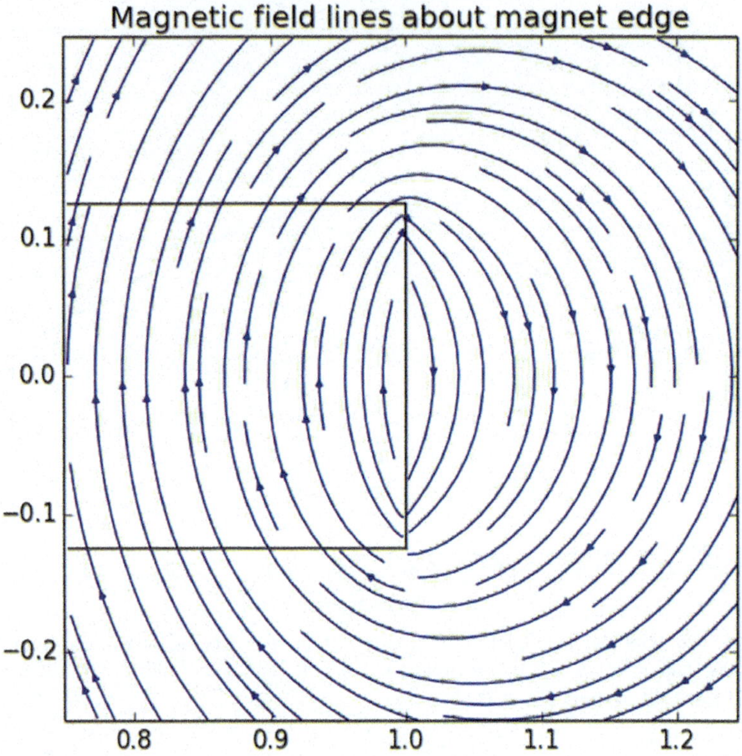


**Fig. 2** Magnetic field lines about the magnet edge, when $a = 1.$ and $H = 0.25$. The abscissa (ordinate) is $R/a$ $(z/a)$. The magnet is shown by the straight lines

The field lines are discontinuous in gradient on the vertical sides of the magnet. This is well known from experiments, which apply iron filings about the boundary of a fixed magnet.

**The cylinder R = a**. The terms in $B_z$ involving $\Pi(n|k)$ in (12) are discontinous at $R = a$, while $B_R$ is continuous across $R = a$. We have about $R = a$,

$$\sqrt{n - k_\pm^2} = \frac{|z \pm \frac{H}{2}|}{r_\pm}; \quad \sqrt{1-n} = \frac{|R-a|}{R+a}; \quad \Pi(n|k_\pm) = \frac{\pi r_\pm (R+a)}{2|z \pm \frac{H}{2}||R-a|} \tag{33}$$

$$B_z(a^\pm, z) = \frac{\mu_0 M \mathrm{sgn}(R-a)}{4}\left[\mathrm{sgn}(z - \frac{H}{2}) - \mathrm{sgn}(z + \frac{H}{2})\right] + \tag{34}$$

$$\frac{\mu_0 M}{2\pi}\left[\frac{(z + \frac{H}{2})K(k_+)}{r_+} - \frac{(z - \frac{H}{2})K(k_-)}{r_-}\right]. \tag{35}$$

The discontinuous terms in (35) only arise when $|z| < H/2$, when

$$\frac{\mu_0 M \text{sgn}(R-a)}{4}\left[\text{sgn}(z-\frac{H}{2}) - \text{sgn}(z+\frac{H}{2})\right] = \frac{\mu_0 M \text{sgn}(a-R)}{2} \tag{36}$$

giving the required jump in $B_z$ of $-\mu_0 M$ on moving outwards across the magnet boundary. When $|z| > H/2$ and $R = a$, the potentially discontinous terms in (35) cancel, so that the only discontinuities in the magnetic field occur across the vertical sides of the cylinder.

**The far field**. In the limit that $|\mathbf{x} - \mathbf{x}_m|$ is much greater than both $a$ and $H$, the magnetic field approaches that of a point dipole field,

$$\mathbf{B} = \frac{\mu_0}{4\pi r_m^3}[3(\mathbf{m} \cdot \mathbf{e}_r)\mathbf{e}_r - \mathbf{m}] \tag{37}$$

where $\mathbf{m}$ is the magnetic moment of the point dipole, $r_m = |\mathbf{x} - \mathbf{x}_m|$ and $\mathbf{x}_m = (x_m, y_m, z_m)$ is the mid point of the point magnet. For $\mathbf{m} = |\mathbf{m}|\mathbf{e}_z$,

$$\mathbf{B} = \frac{\mu_0|\mathbf{m}|}{4\pi r_m^5}\left(3(z-z_m)(x-x_m), 3(z-z_m)(y-y_m), 3(z-z_m)^2 - r_m^2\right) \tag{38}$$

Since the magnetisation vector $\mathbf{M}$ represents magnetic moment per unit volume, $\mathbf{m} = \pi a^2 H\mathbf{M}$. Consequently, the far field appears to arise from a circular current source $I$ around the magnet's curved boundary,

$$I = MH \tag{39}$$

and far field distances scale naturally with the cube root of the magnet volume.

## 4 Magnetised Disk

Neodymium cylindrical or disk magnets typically have $H/a < 0.5$, and it is not uncommon for $H/a$ to be as small as 0.25. Then if $R \simeq a$, the parameter $k_0^2$ in (16) differs by only a few percent from the limiting case of $H = 0$, suggesting we consider the idealised disk example of $H/a \searrow 0$.

From (1), the vertical component of $\mathbf{B}$ from an idealised cylindrical disk is

$$B_z(R, z) = \frac{\mu_0 MaH}{4\pi}\int_0^{2\pi}\frac{d\phi\,(a - R\cos\phi)}{\left[R^2 + z^2 + a^2 - 2aR\cos\phi\right]^{\frac{3}{2}}} \tag{40}$$

$$= \frac{\mu_0 MH}{2\pi\sqrt{(R+a)^2 + z^2}}\left[K(k) - \frac{(R^2 + z^2 - a^2)}{((R-a)^2 + z^2)}E(k)\right] \tag{41}$$

$$k^2 = \frac{4aR}{\left[(R+a)^2 + z^2\right]}. \tag{42}$$

At the centre of the disk magnet, $B_z(0,0) = \mu_0 MH/(2a)$, consistent with (27). In the far field, where $R >> a$ and $z >> a$, $B_z \simeq \mu_0 m(3z^2 - r^2)/(4\pi r^5)$ where $r^2 = R^2 + z^2$ and $m$ is the dipole moment $\pi a^2 HM$, agreeing with (38).

The disk model breaks down around $R = a$ and $z = 0$, where $k$ in (42) equals unity, and so $K(k)$ is unbounded there. The expression keeping $k_0$ away from unity in (16) depends on $H^2/a^2$, which we have ignored in this section.

From (23) or (25), $B_R$ from an idealised cylindrical disk is

$$B_R(\mathbf{x}) = \frac{\mu_0 MHz}{2\pi R\sqrt{(R+a)^2 + z^2}} \left[\frac{(R^2 + a^2 + z^2)}{((R-a)^2 + z^2)} E(k) - K(k)\right] \tag{43}$$

where $k$ is defined in (42). In the far field where $R >> a$ and $z >> a$, $B_R \simeq 3\mu_0 mRz/(4\pi r^5)$, in agreement with (38).

The results in (40) and (43) show that the magnetic field components from a cylindrical disk magnet are proportional to the height $H$ of the magnet. The independent variables $R$ and $z$ enter as $R/a$ and $z/a$, showing that there is a self-similar field $\mathbf{b}$, whose spatial variation scales as the radius of the disk $a$.

$$\mathbf{B} = \frac{\mu_0 MH}{2a}\mathbf{b}\left(\frac{R}{a}, \frac{z}{a}\right); \quad \mathbf{b}(0,0) = 1. \tag{44}$$

The field lines for the non-dimensional magnetic field $\mathbf{b}$ in (44) are plotted in Fig. 3, where distances are plotted relative to the magnet radius, $a$.

## 5 Summary

The mathematical model developed here assumes uniform, unidirectional magnetisation, aligned along the axis of a finite cylinder. The magnetic field has only axial and radial components in a cylindrical coordinate system. The axial (radial) field $B_z$ ($B_R$) is described in terms of complete elliptic integrals of the first and third (second) kinds. There is one elliptic characteristic and two elliptic parameters. The vertical magnetic field component $B_z$ is bounded everywhere, but $B_R$ is logarithmically divergent about the upper and lower magnet edges.

The elliptic integrals collapse to algebraic expressions, along the axis $R = 0$. The shape of the magnetic field lines depends on the ratio of lengths $H/a$, while the strength of $\mathbf{B}$ depends linearly on the magnitude of the magnetisation, $|\mathbf{M}|$. The magnetic field is axial along the axis.

The radial component of the magnetic field $B_R$ is zero on the transverse plane $z = 0$ through the centre of the magnet. The fixed points (where $\mathbf{B} = \mathbf{0}$) are associated with the circle $R = a$ on the surface of the magnet, where $z = 0$. However, these

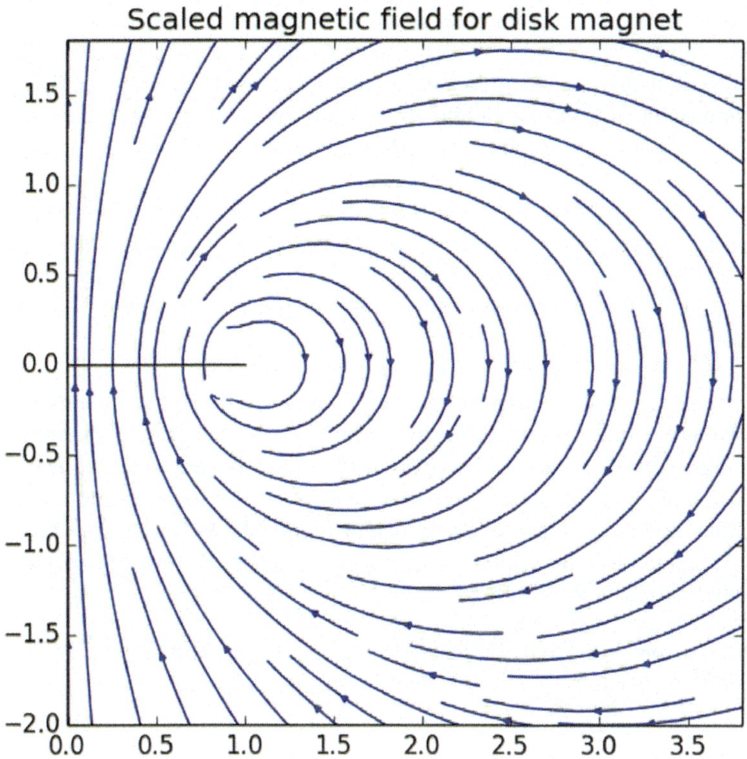


**Fig. 3** Field lines for the non-dimensional magnetic field **b** in (44). The abscissa (ordinate) is $R/a$ ($z/a$). The disk position is the horizontal line

points occur where $B_z$ is discontinuous, so there is ambiguity about the meaning of the fixed points for the magnetic field.

The results above support a conceptual model, comprising of three elements, for the magnetic field from commercially available cylindrical neodymium magnets, in which their height $H$ is somewhat smaller than their radius $a$. Firstly, away from the magnet, the magnetic field is approximately that from a disk magnet, whose magnitude depends linearly on the height $H$ of the magnet, and whose spatial variation scales with the disk radius, as shown in (44).

Secondly, the coaxial circle $(R, z) = (a, 0)$ is associated with the fixed points of $\mathbf{B}$. On this circle, some of the complete elliptic integrals are singular, and $B_z$ is discontinuous.

Thirdly, the magnetic field about the upper and lower magnet surfaces depends sensitively on both the smoothness of the magnet surface, and the uniformity of magnetisation of magnetic domains. Even for uni-directional magnetisation, the magnetic field can completely reverse about the magnet surface, for moderate variations in the magnitude of domain magnetisation.

Cylindrical neodymium magnets are the strongest permanent cylindrical magnets commercially available and are widely used in industrial applications. Their magnetisation results from magnetic domains about $10^{-5}$m in size. Commercially available neodymium magnets have small height to width ratios. We expect the corresponding ratio for magnetic domains to be much greater. If local magnetisation is fairly variable across surface magnetic domains, then we expect that many field reversals and enhancements will occur about the surfaces of a neodymium magnet, over the length scales of the magnetic domains, producing a highly variable surface magnetic field. This effect should be greatest about the surfaces, edges and vertices of magnets, where magnetic domain pinning will be weakest, and variability of magnetisation will be greatest.

All magnetic effects can be imagined to result from the movement of electrical charges, leading to the concept of Amperian currents, which generate the magnetic field. We have used this concept in this paper, by assuming that the magnetic field results from surface currents. This approach has the advantage that the **H** field need not be considered but forces the transverse component of **B** to be discontinuous across the magnet boundary. A novel aspect of this paper has been deriving the complete elliptic integral of the third kind, which arises from the constant discontinuity in $B_\theta$ across the vertical sides of the cylinder.

**Acknowledgements** This work was funded by the New Zealand M.B.I.E. contract UOAX1711 *Development of IPT roadway charging systems.*

## References

Byrd PF, Friedman MD (1954) Handbook of elliptic integrals for engineers and physicts. Springer, Berlin

Cao Q, Han X, Li L (2014) Configurations and control of magnetic fields for manipulating magnetic particles in microfluidic applications: magnet systems and manipulation mechanisms. Lab Chip 14(15):2762–2777

Corson D, Lorrain P (1962) Introduction to electromagnetic fields and waves. W.H. Freeman, San Francisco

Folks L, Woodward RC (1998) The use of MFM for investigating domain structures in modern permanent magnet materials. J Magn Magn Mater 190(1–2):28–41

Göddenhenrich TH, Hartmann U, Anders M, Heiden C (1988) Investigation of Bloch wall fine structures by magnetic force microscopy. J Microsc 152(2):527–536

Goncharov AA, Brown IG (2007) Plasma devices based on the plasma lens-a review of results and applications. IEEE Trans Plasma Sci 35(4):986–991

Gradshteyn IS, Ryzhik IM (1980) Tables of integrals, series, and products, 4th edn. Academic, New York

Grant KM, Hemmert JW, White HS (2002) Magnetic field-controlled microfluidic transport. J Am Chem Soc 124(3):462–467

Hartmann U (1999) Magnetic force microscopy. Annu Rev Mater Sci 29(1):53–87

Helander PD, Drevlak M, Zarnstorff M, Cowley SC (2020) Stellarators with permanent magnets. Phys. Rev. Lett. 124(9):095001

Kaye GWC, Laby TH (1995) Table of physical and chemical constants, 16th edn. Longman

Melone JJ, Ledingham KWD, McCanny T, Burris-Mog T, Schramm U, Grötzschel R, Akhmadaliev S, Hanf D, Spohr KM, Bussmann M (2012) Characterisation of permanent magnetic quadrupoles for focussing proton beams. Nucl Instrum Methods Phys Res Sect A: Accel, Spectrometers, Detect Assoc Equip 676:126–134

Nevcas D, Klapetek P, Neu V, Havlivcek M, Puttock R, Kazakova O, Hu X, Zajivckova L (2019) Determination of tip transfer function for quantitative MFM using frequency domain filtering and least squares method. Sci Rep 9(1):1–15

Sattel D, Reid J (2006) Modelling of airborne EM anomalies with magnetic and electric dipoles buried in a layered earth. Exp Geophys 37(3):254–260

Schwarz A, Wiesendanger R (2008) Magnetic sensitive force microscopy. Nano Today 3(1–2):28–29

Shaar R, Feinberg JM (2013) Rock magnetic properties of dendrites: insights from MFM imaging and implications for paleomagnetic studies. Geochem Geophys Geosyst 14(2):407–421

Stratton JA (1941) Electromagnetic theory. McGraw-Hill, New York

Vacquier V, Steenland NC, Henderson RG, Zietz I (1951) Interpretation of aeromagnetic maps. Geol Soc Am. Mem 47:151

Weir GJ, Chisholm G, Leveneur J (2020) The magnetic field about a three-dimensional block neodymium magnet, in press. ANZIAM J 62(4):386–405

Whittaker IET, Watson EN (1927) A course in modern analysis. Cambridge University Press, Cambridge

Won C, Wu YZ, Zhao HW, Scholl A, Doran A, Kim W, Owens TL, Jin XF, Qiu ZQ (2005) Studies of Fe Mn/ Co/ Cu (001) films using photoemission electron microscopy and surface magneto-optic Kerr effect. Phys. Rev. B 71(2):024406

Yuan QL, Wang YZ, Shi CT, Chen GX, Zhang SX, Chen RHH, Liu MS (1987) Application of permanent magnet lenses in accelerators beam transport and focusing. Chinese Phys C 11(1):11–15

# New Models for Deformations: Linear Distortion and the Failure of Rank-One Convexity

**Seyed Mohsen Hashemi and Gaven J. Martin**

## 1 Introduction

### *1.1 Aspects of Material Science*

The use of nonlinear elasticity to describe martensitic transformations and their microstructure starts with Ball and James (1987), following the modelling approach of J.L. Ericksen for elastic crystals. According to Bhattacharya (2003), the Martensitic transition is a diffusionless structural transition from a high symmetry to a low symmetry crystallographic phase. The lecture notes of Ball (Défauts dans les cristaux et dans les cristaux liquides) give a very good overview of the subject at an introductory level. It is in fact one of the most studied phase transitions in material science. This transition is often induced by changing temperature or stress and shows athermal character and proceeds intermittently as a sequence of avalanches ultimately producing a very complex multiscale microstructure.

Recently, the theory of mappings of finite distortion has emerged and shown promise in modelling various aspects of nonlinear materials science with interesting associated extremal problems (see for example, Astala et al. 2009, 2005; Hencl and Koskela 2014; Iwaniec et al. 2013; Iwaniec and Martin 2001; Iwaniec et al. 2014; Iwaniec and Onninen 2016, 2010; Martin and Yao 2020). Many of the distortion functionals studied in these cases are lower semicontinuous and extremals are quite regular—typically diffeomorphisms. However, experimental evidence of fine pattern formation (see Bhattacharya 2003) suggests candidate energy models for the Austenite-to-Martensite transition should lack lower semi-continuity, allowing for

Gaven J. Martin—Work of both authors is partially supported by the New Zealand Marsden Fund. We are extremely grateful for the careful reading of an earlier version of this paper by an anonymous referee whose suggestions significantly improved the exposition of this article.

S. M. Hashemi · G. J. Martin (✉)
N.Z. Institute for Advanced Study, Massey University, Auckland, New Zealand
e-mail: g.j.martin@massey.ac.nz

S. M. Hashemi
e-mail: S.M.Hashemi@massey.ac.nz

 
R. McKibbin et al. (eds.), *Proceedings of the Forum "Math-for-Industry" 2019*,
Mathematics for Industry 36, https://doi.org/10.1007/978-981-19-1154-5_4

**Fig. 1** These images are from Chu and James's experiments in Cu-Al-Ni single crystals, and Ball (Défauts dans les cristaux et dans les cristaux liquides, lecture notes). Reproduced with permission of authors

the study of minimising sequences rather than minimisers. The well-known images in Fig. 1 reinforce the case for the lack of higher regularity for phase transitions in these sorts of structural transformations.

Here we suggest and analyse an "energy" related to the linear distortion of a deformation. Here polyconvexity and lower semi-continuity are lost. We thereby expect to gain models exhibiting non-smooth extrema, or more precisely minimising sequences with fine-scale structures and low regularity. Distortion functionals are polyconvex in two dimensions, so it is unlikely we will see these complicated microscale structures with purely two-dimensional models only using distortion. However, an interesting feature of our model is that although fundamentally three dimensional, limit regimes yield two-dimensional and even one-dimensional models.

## *1.2 Distortion Functions*

Distortion functionals are scale-invariant measures of the anisotropic nature of a deformation. We view the $L^p$-norm, $1 \leq p \leq \infty$ of the linear distortion of a deformation as a stored energy functional, and use the term "$p$-conformal energy". All these conformal energy functionals are measures of the deviation from a conformal mapping, and in the problems we consider, should a conformal mapping be a candidate, it will be an absolute minimiser. For the purposes of this article, we restrict to three dimensions and consider deforming a body $\Omega \subset \mathbb{R}^3$ by a **homeomorphism** $f : \Omega \to f(\Omega) \subset \mathbb{R}^3$. Typically, the assumption that a deformation is a homeomorphism is given to us by the principle of non-interpenetrability of matter, see Ball (1981). It is a remarkable feature of mappings of bounded distortion, coming from modulus of continuity estimates on mappings and their inverses, that this topological condition is retained under limits. Let $\Omega \subset \mathbb{R}^n$, $n \geq 3$, be a domain and $f : \Omega \longrightarrow \mathbb{R}^n$ be a homeomorphism. Typically, one assumes some regularity of the deformations in question, namely,

$$f \in W^{1,3}_{loc}(\Omega),$$

the Sobolev space of mappings whose first derivatives are in $L^3$. This condition is basically the minimal assumption one can make to ensure that the Jacobian determinant is locally integrable and that something like the change of variables formula might hold (Iwaniec and Martin 2001).

### 1.3 Polyconvex Distortion Functions

The distortion inequality assumes that the full differential of a mapping is controlled by the Jacobian; there is a constant $1 \leq K < \infty$ such that

$$|Df(x)|^2 \leq K J(x, f), \qquad \text{almost every} x \in \Omega. \tag{1}$$

Such mappings are called **quasiconformal**. The norm here is typically taken to be the operator norm,

$$|Df(x)| = \sup_{|h|=1} |Df(x)h|. \tag{2}$$

Any other equivalent norm—for instance the Hilbert-Schmidt norm

$$\|Df(x)\|^2 = \sum_{i,j} \left(\frac{\partial f^i}{\partial x_j}\right)^2$$

will give the same class of mappings though the exact constant may vary. Deformations satisfying the distortion inequality enjoy many properties, including higher regularity

$$f \in W^{1,3+\epsilon(K)}_{loc},$$

positive Jacobian $J(x, f) > 0$ almost everywhere and the change of variables formula (Iwaniec and Martin 2001). Thus, one can define the measurable distortion function

$$\mathbf{K}(x, f) = \frac{\|Df(x)\|^2}{J(x, f)}. \tag{3}$$

It is not obvious but not too hard to see that $\mathbf{K}(x, f)$ is a convex function of the minors of $Df(x)$—the function $(x, y) \mapsto \frac{x^2}{y}$ is convex on the positive quadrant of $\mathbb{R}^2$. Thus one can consider various extremal problems, for instance, minimising the $L^p$-norms of $\mathbf{K}(x, f)$ among deformations with prescribed boundary values linking with the calculus of variations. Some novel phenomena arise in such problems including the Nitsche phenomena (Astala et al. 2005; Martin and McKubre-Jordens 2012) where geometric obstructions such as conformal modulus preclude the existence of a minimiser. The connection with the Nitsche phenomena for harmonic mappings is through the remarkable formula

$$\int \mathbf{K}(x, f)\, dx = \int \|Dg(w)\|^2\, dw, \tag{4}$$

which holds for suitably regular Sobolev homeomorphisms $f$ with inverse $g = f^{-1}$, Astala et al. (2009). It is expected that homeomorphic minimisers are smooth when they exist (and they should exist for all $p > 1$) if there is a candidate "barrier".

## 1.4 Definition of Linear Distortion

For $x \in \Omega$, and $f : \Omega \to \mathbb{R}^3$ a homeomorphism, we set

$$\mathcal{H}(x, f) = \limsup_{r \to 0} \frac{\max_{|h|=r} |f(x+h) - f(x)|}{\min_{|h|=r} |f(x+h) - f(x)|}, \tag{5}$$

and then define $\mathcal{H}(f) = \text{ess sup}\{\mathcal{H}(x, f) : x \in \Omega\}$.

If $\mathcal{H}(x, f)$ is bounded in $\Omega$, then $f$ is $K$-quasiconformal for some $K$ bounded by a function of $\mathcal{H}$. In particular $f$ lies in the Sobolev space $W^{1,3}_{loc}(\Omega)$ and satisfies another equivalent form of the distortion inequality,

$$\max_{|\zeta|=1} |Df(x)\zeta| \le \mathcal{H}(f) \min_{|\zeta|=1} |Df(x)\zeta|.$$

Geometrically, this means that the differential $Df(x) : \mathbb{R}^n \longrightarrow \mathbb{R}^n$ maps the unit sphere to an ellipsoid of bounded eccentricity.

It is worth observing a remarkable theorem of Hencl and Koskela (2006), which shows that one only requires that the **liminf** in (5) is bounded to find the limsup is also bounded and therefore the deformation is quasiconformal (with the same constants).

**A natural question is: *Can scale-invariant problems, and among them the formation of Martensitic microstructure, be modelled by distortion functionals?***

A connection between conformal energy functionals and the formation of microstructures is predicated in the belief that at molecular scales we see "infinitesimal" linear deformations of uniformly bounded distortion, as illustrated below in Fig. 2.

There are many outstanding issues that remain as to how interfacial energies could be incorporated so as to cause the emergence of length scales and solve boundary value problems. Examples of these can be seen in Ball and Carstensen (1991, Fig. 1) where potential nonclassical interfaces are presented. In that figure (b) has now been observed by Seiner (J. Ball personal communication).

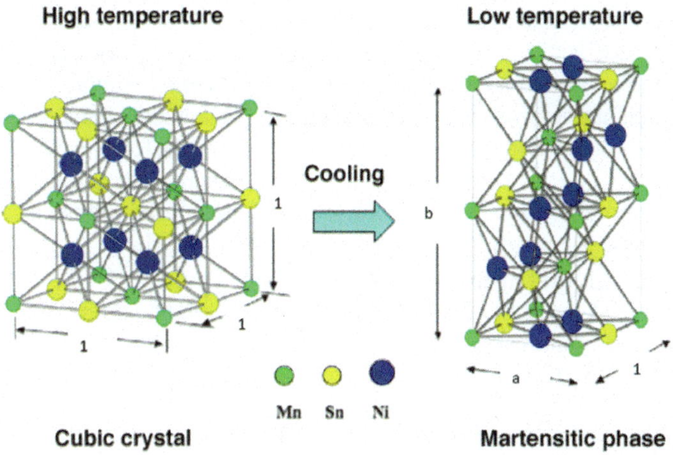


**Fig. 2** Martensitic phase after cooling down process

## 2 A Model Problem

Let $\mathbf{Q} = [0, 1]^3$ be the unit cube in $\mathbb{R}^3$ and $A \in GL(3, \mathbb{R})$. Consider the problem:

$$\text{minimize}: \int_{\mathbf{Q}} \Phi[\mathcal{K}(x, f)]\, dx \quad \Big| \quad f : \mathbf{Q} \longrightarrow \mathbb{R}^n \;\; \text{is quasiconformal.} \tag{6}$$

Here $\Phi$ is convex increasing and $\mathcal{K}(x, f)$ is a distortion function. We want to discuss boundary constraints. Thus denote the six faces of $\mathbf{Q}$ as $\sigma_j$, $j = 1, 2, \ldots, 6$. We consider imposing the following possible constraints.

- $f|\partial\mathbf{Q} = A$; in this case, $f$ is assumed pointwise equal to $A$ on the boundary. Typically the linear map is the minimiser, but it is not unique for linear distortion unless $A$ has repeated singular values. This failure of uniqueness occurs as one can perturb an intermediate singular value without affection the linear distortion. For instance $(x, y, z) \mapsto (x, 2 + \phi(y), 3z)$ has linear distortion three and is linear on the boundary if $\phi$ is compactly supported with small gradient.
- $f(\sigma_j) = A(\sigma_j)$, in this case, we have faces being mapped to faces and typically a linear minimiser; or
- $f(\sigma_j) \subset N_\epsilon(A(\sigma_j))$, $j = 1, \ldots, n$; in this case, we have faces being mapped within an $\epsilon$ neighbourhood of faces. Then the linear map is never a minimiser for linear distortion; or
- $\partial f(\mathbf{Q}) \subset N_\epsilon(A(\partial\mathbf{Q})$ as above.

For the last two, we take a limit as $\epsilon \to 0$. In all cases, the linear map is the minimiser for $\|\mathbf{K}(x, f)\|_p$, $1 \le p \le \infty$, Astala et al. (2005). In the last two cases, the linear map may be the limit of a minimising sequence but is not a minimum.

Notice that quasiconvexity of the functional at (6) would guarantee that the minimum should be attained by the linear map $x \mapsto Ax$. In what follows, we simply write $\mathcal{H}(A)$ for $\mathcal{H}(x, x_0 + Ax)$ as the linear distortion of an affine mapping is constant.

One of our main results whose proof we sketch here is the following.

**Theorem 1** (Linear maps are not minimizers) *Let $\Omega \subset \mathbb{R}^n$ be a bounded domain and $A \in GL(3, \mathbb{R})$ with three distinct singular values. Then there is $\delta > 0$ with the following property:*
*For every $\epsilon > 0$, there is a quasiconformal mapping $f_\epsilon : \Omega \to \mathbb{R}^n$ so that*

1. $\|f_\epsilon(z) - A(z)\|_{L^\infty(\Omega)} < \epsilon$.
2. $$\int_{\mathbf{Q}} \Phi[\mathcal{H}(x, f_\epsilon)] < \int_{\mathbf{Q}} \Phi[\mathcal{H}(A)] - \delta.$$

*Here $\mathcal{H}(x, f_\epsilon)$ is the linear distortion of $f_\epsilon$.*

This result is true in $\mathbb{R}^n$, $n \ge 3$ as soon as there are three distinct singular values for $A$. This result exhibits the failure of lower semi-continuity.

An intriguing question is how big can the jump be? We shall return to this question in Sect. 5.

## 3 The Linear Distortion

For many years—perhaps since the inception of the higher dimensional theory of quasiconformal mappings—it was assumed that, like other measures of distortion for mappings such as the inner and outer distortion, the linear distortion was lower semi-continuous; that is *the distortion of a limit is never more than the limit of the distortions*. T. Iwaniec gave a striking example that refuted this belief in 1998, Iwaniec (1998). The key element of his construction is that the linear distortion function fails to be rank-one convex in dimension higher than 2.

**Theorem 2** (Iwaniec 1998) *For each dimension $n \ge 3$ and dilatation $H > 1$, there exists a sequence $\{f_j\}_{j=1}^{\infty}$ of $H$-quasiconformal mappings of $\mathbb{R}^n$ converging uniformly to a quasiconformal linear map $f : \mathbb{R}^n \longrightarrow \mathbb{R}^n$ whose dilatation is greater than $H$.*

The dimension is greater than two and is necessary of course. Understanding the deep connections between problems of semi-continuity, the various notions of convexity of functionals and material science, Iwaniec realised the question could be reduced to disproving the rank-one convexity of the linear distortion functional. He gave a specific family of examples establishing this theorem. Our results show that this failure is a generic feature and in fact the limit can increase by a factor of $\sqrt{2}$, see Theorem 4 below. We now discuss the proof of these and related results.

### 3.1 Rank-One Convexity

A function $\mathcal{F} : U \subset \mathbb{R}^{n\times n} \longrightarrow \mathbb{R}$, defined on an open set $U$ of $n \times n$ matrices, is said to be **rank-one convex** at $A_0 \in U$ if for every rank-one matrix $X \in \mathbb{R}^{n\times n}$ the function

$$t \mapsto \mathcal{F}(A_0 + tX)$$

is convex near $t = 0$. This is a weaker assertion than that classically requiring convexity for all $t > 0$, Dacorogna (2009), but the reader will see is immaterial for our considerations. Recall that $X$ has rank one if and only if it is the tensor product of two vectors $X = u \otimes v$, where $u, v \in \mathbb{R}^n$.

The Hadamard jump condition asserts that the piecewise linear function $F = \{A, B\}$ is continuous across an interface if and only if $A - B = u \otimes v$, and the normal to the interface is $v$ (Fig. 3).

It is an exercise to show that the linear distortion functional $\mathcal{H} : GL^+(3, \mathbb{R}) \to \mathbb{R}$, where we identify $GL^+(3, \mathbb{R})$ as linear homeomorphisms, is defined by

$$\mathcal{H}(A) = \mathcal{H}(x, A) = \frac{\max_{|\zeta|=1} |A\zeta|}{\min_{|\zeta|=1} |A\zeta|} = \sqrt{\frac{\lambda_3}{\lambda_1}},$$

where $\lambda_1 \leq \lambda_2 \leq \lambda_3$ are the ordered eigenvalues of $A^t A$. Iwaniec showed (by hand!) that $\mathcal{H}$ is not rank-one convex at $A_c$ for $c > 1$, where

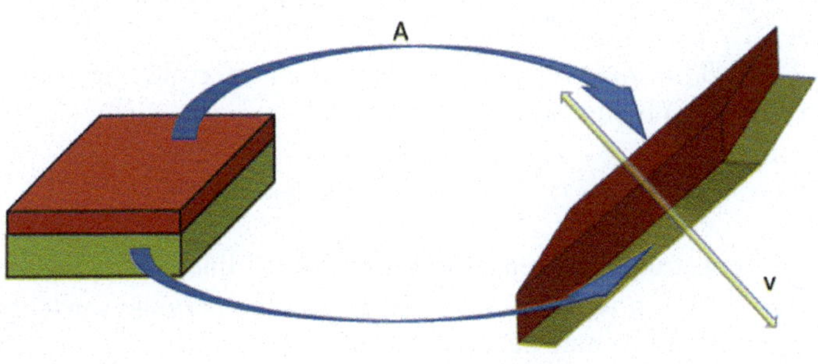


**Fig. 3** Rank-one connected affine mappings with a planar interface

$$A_c = \begin{bmatrix} 1 & 0 & 0 \\ 0 & c & 0 \\ 0 & 0 & c^2 \end{bmatrix}.$$

Note $\mathcal{H}(A_c) = c^2$. He chose two vectors $u = (1, b, c)$ and $v = (1, -b, c)$. If $X = u \otimes v$, then he proves that

$$\mathcal{H}(A + t_+X) = \mathcal{H}(A + t_-X) = \sqrt{\frac{\lambda_{max}}{\lambda_{min}}} = c^2 - 2t^2 + \text{higher powers of } t,$$

where $\lambda_{max}$ and $\lambda_{min}$ are the largest and the smallest eigenvalues of matrix $(A + tX)^T(A + tX)$. It becomes obvious that $\mathcal{H}(A + tX) < \mathcal{H}(A)$ for $t$ sufficiently small.

## 3.2 Optimal Rank-One Directions

Our next result gives the best rank one direction to deform a linear mapping. We only sketch the ideas for the proof of this result here, mostly in Sect. 4. Complete details will appear in M. Hashemi's PhD Thesis.

**Theorem 3** *Given $A \in GL^+(3, \mathbb{R})$ with distinct singular values. Then up to sign, there is a unique rank one matrix $B_0 = u_0 \otimes v_0$, $\|u_0\| = \|v_0\| = 1$, with the following properties.*

- $\frac{d}{dt}\Big|_{t=0} \mathcal{H}(A + tB_0) = 0, \quad \text{and} \quad \frac{d^2}{dt^2}\Big|_{t=0} \mathcal{H}(A + tB_0) < 0,$
- $\frac{d^2}{dt^2}\Big|_{t=0} \mathcal{H}(A + tB_0) \leq \frac{d^2}{dt^2}\Big|_{t=0} \mathcal{H}(A + tB),$

*for all $B = u \otimes v$, $\|u\| = \|v\| = 1$, with $\frac{d}{dt}\Big|_{t=0} \mathcal{H}(A + tB) = 0$*

Before discussing the proof of Theorem 3, we show how it can be used to establish Theorem 1.

Given $A$ and $B_0 = u_0 \otimes v_0$ as above, there is an interval $[t_-, t_+]$ on which the function $\mathcal{H}(A + tB_0)$ is smooth and concave, and

$$\mathcal{H}(A + t_-B_0) \ \& \ \mathcal{H}(A + t_+B_0) < \mathcal{H}(A)$$

. One can now construct a sequence converging to the linear mapping with strictly smaller linear distortion as follows. First construct the periodic sawtooth function $a(r)$ whose graph we illustrate below (Fig. 4).

Then define function $f_j : \mathbb{R}^n \longrightarrow \mathbb{R}^n$ with equation

$$f_j(x) = Ax + \frac{1}{j}\, a(j\mathbf{u} \cdot x)\mathbf{v},$$

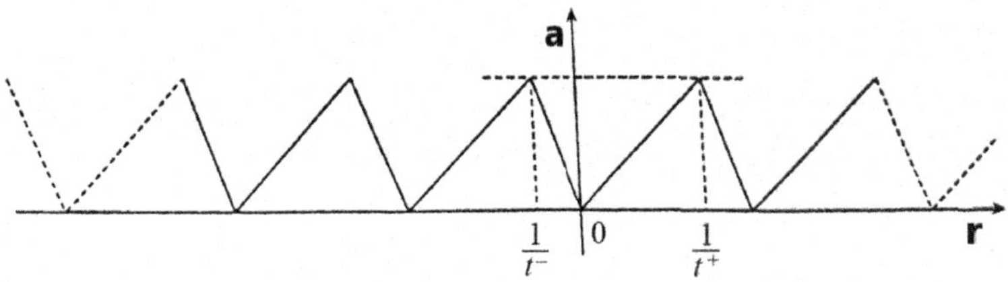


**Fig. 4** The sawtooth function $a(r)$

where $j = 1, 2, 3, \ldots m$. and $a$ is a piecewise linear function on $\mathbb{R}$ such that

$$a(r) = \begin{cases} t^- r & \text{if } \frac{1}{t^-} \leq r \leq 0 \\ t^+ r & \text{if } 0 \leq r \leq \frac{1}{t^+} \end{cases}$$

and then extended periodically to the real line. The functions $f_j(x)$ have differential that assumes only two values, which are independent of $j$:

$$Df_j(x) = A + a'(j u_0 \cdot x)\, u_0 \otimes v_0 \in \{A + t_\pm B_0\}$$

. It is easy to see that for a deformation $F : \Omega \to \mathbb{R}^3$, which has nonsingular derivative at a point $x_0 \in \Omega$ we have $\mathcal{H}(x_0, F) = \mathcal{H}(DF(x_0))$. Thus for all $j \geq 1$ and almost every $x$ we have

$$\mathcal{H}(x, f_j) = \mathcal{H}(Df_j(x)) \leq \max\left\{\mathcal{H}(A + t_\pm B_0)\right\} < \mathcal{H}(A).$$

Effectively, the piecewise affine mappings $f_j$ oscillate at finer and finer scales and in the limit converge uniformly to $f(x) = Ax$ while their linear distortion stays a definite amount below that of $A$ (Fig. 5).

To address the size of the energy deficiency identified in Theorem 1, we will find the maximum interval on which $\mathcal{H}(A + tB_0)$ is smooth and concave. To do this, we find the discriminant of the eigenvalue equation for $\mathcal{H}(A + tB_0)$ and determine its roots, as it is the transverse crossing of the eigenvalues where $\mathcal{H}$ loses smoothness.

## 4 Optimal Directions

Let $A \in GL^+(3)$ be a nonsingular matrix. We write $A \in Sing(1, \alpha, \beta)$ if the ordered singular values of $A$ are $1, \alpha$ and $\beta$. That is the ordered eigenvalues of $A^T A$ are $1 \leq \alpha^2 \leq \beta^2$. For any nonsingular matrix $A$, the assumption that 1 is a singular value can be simply achieved by scaling. This scaling does not change the linear distortion.

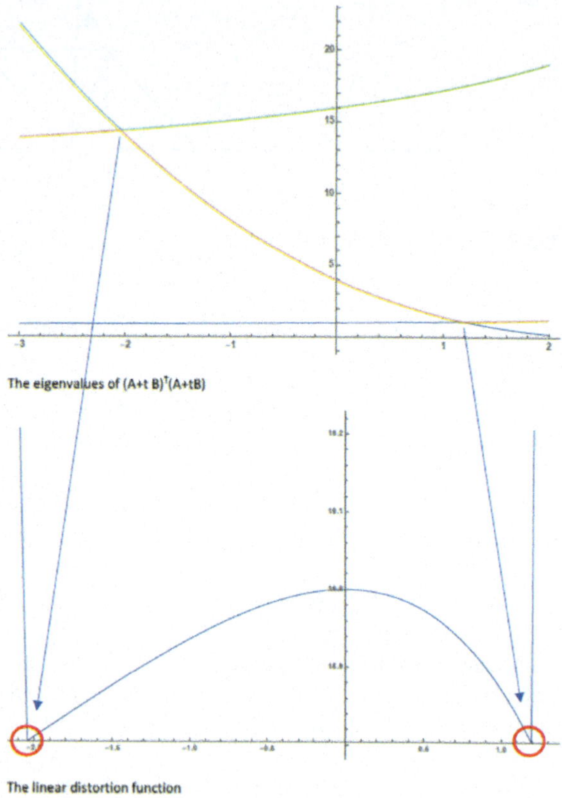


**Fig. 5** $\mathcal{H}$ is concave in a neighbourhood of 0 determined by the eigenvalue crossings

To prove Theorem 3, we suppose we are given $A \in Sing(1, \alpha, \beta)$. We want to determine a rank one matrix $B_0 = u_0^T v_0$ for which the following holds.

$$\left.\frac{d}{dt}\right|_{t=0} \mathcal{H}(A + tB_0) = 0 \tag{7}$$

and for all other rank one matrices $B$ with $\left.\frac{d}{dt}\right|_{t=0} \mathcal{H}(A + tB) = 0$ we have

$$\left.\frac{d^2}{dt^2}\right|_{t=0} \mathcal{H}(A + tB_0) \leq \left.\frac{d^2}{dt^2}\right|_{t=0} \mathcal{H}(A + tB). \tag{8}$$

To establish (8), we require the quadratic term in the series for $\mathcal{H}(A + tB)$ to be as large and negative as possible and to identify the unit vectors $u_0$ and $v_0$ which achieve this. A calculation reveals that after a conformal conjugacy we may reduce to the case that $A$ is a diagonal matrix. Then for $0 \leq r, s \leq 1$ we put

$$u^T = (\sqrt{1-r^2}, r\cos(\theta_1), r\sin(\theta_1)), \tag{9}$$
$$v^T = (\sqrt{1-s^2}, s\cos(\theta_2), s\sin(\theta_2)) \tag{10}$$

and then put $B = u^T.v$. Set $X = (A + tB)^T(A + tB)$ and calculate the Taylor series of the eigenvalues of $X$ to second order.

$$\lambda_{min} = 1 + \mu_1 t + \mu_2 t^2 + O(t^3)$$
$$\lambda_{max} = \beta^2 + \eta_1 t + \eta_2 t^2 + O(t^3).$$

Up to $O\left(t^3\right)$,

$$\mathcal{H}(A + tB) = \beta + t\frac{\left(\eta_1 - \beta^2\mu_1\right)}{2\beta} - \frac{t^2\left(\beta^4\left(4\mu_2 - 3\mu_1^2\right) + 2\beta^2\eta_1\mu_1 - 4\beta^2\eta_2 + \eta_1^2\right)}{8\beta^3}.$$

Now the coefficients $\mu_i, \eta_i, i = 1, 2,$ of $\lambda_{max}$ and $\lambda_{min}$ are polynomials in the four variables $r$, $s$, $\sin(\theta_1)$, $\sin(\theta_2)$. Next (7) gives us the constraint

$$0 = \eta_1 - \beta^2\mu_1.$$

Subject to this constraint we want to maximise

$$\mathcal{Q} = \eta_2 - \beta^2\mu_2.$$

We use Lagrange multipliers to identify the extreme value of $\mathcal{Q}$. From the system of equations so obtained, we eliminate the variables $r$ and $s$. This leads to a new minimisation problem for the trigonometric algebraic function $\mathcal{Q}$ when written terms of the two variables $\theta_1$ and $\theta_2$ which lie in $[0, \pi]$, once we define $u$ and $v$ as at (9). The function $\mathcal{Q}$ has various (non-obvious) symmetries that are observed by graphing a few examples and once established we are able to prove the critical points giving a minimum for $\mathcal{Q}$ lie on the line $\theta_2 = \pi - \theta_1$. Returning to the Lagrange multiplier system, we next determine that $r = s$ (Fig. 6).

Now, returning to (9) with $\theta_2 = \pi - \theta_1$ and $r = s$ we can directly compute the minimum value for $\mathcal{Q}$, giving solutions for $u$ and $v$ in terms of the singular values of $A$. We find $u, v =$

$$\frac{1}{\sqrt{2}\sqrt{\alpha^2 + \alpha + \beta^2 + (\alpha - 1)\beta + 1}}\begin{pmatrix} (\beta - 1)/\sqrt{\beta + 1} \\ \pm\sqrt{2\alpha^2 + 2(\beta + 1)\alpha + \beta^2 + 1} \\ (\beta - 1)\sqrt{\beta}/\sqrt{\beta + 1} \end{pmatrix},$$

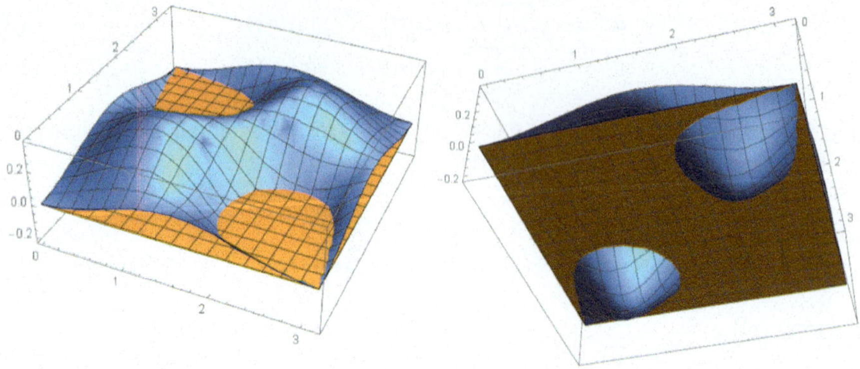

**Fig. 6** Left: The function $\mathcal{Q}(\theta_1, \theta_2)$ when $A = Sing(1, 2, 4)$ on $[0, \pi] \times [0, \pi]$. Right: The minimums of function $\mathcal{Q}$ lie on the line $\theta_2 = \pi - \theta_1$

where $u$ is identified by the choice + in the middle. With these choices of $u$ and $v$ now set

$$B_0 = u^T v. \tag{11}$$

We next compute the first two terms of the series for

$$\sqrt{\frac{\lambda_{max}(t)}{\lambda_{min}(t)}}.$$

To second order,

$$\mathcal{H}(A + tB_0) = \beta - \frac{(\beta-1)^3\beta\, t^2}{4(\alpha+1)(\beta+1)(\alpha+\beta)\left(\alpha^2+(\alpha-1)\beta+\alpha+\beta^2+1\right)}.$$

With this setup $\lambda_{max}$ and $\lambda_{min}$ are increasing, and $\lambda_{mid}$ is decreasing. We have to find the largest $t_- < 0$ and $t_+ > 0$ so the eigenvalues do not cross. For this, we must examine the eigenvalue equation

$$\det[(A + tB_0)^T(A + tB_0) - \lambda^2 Id], \tag{12}$$

and then compute its discriminant. One might think that (12) is cubic in $\lambda^2$ and of degree 6 in $t$. Remarkably, it is quadratic in $t$, a consequence of Jacobi's theorem. We can therefore easily compute the two roots in $t$. We identify one of them as

$$t = \frac{2(\lambda-1)(\alpha+\lambda)(\beta-\lambda)}{r^2(\alpha(2\beta-\lambda-1)+(\beta-2)\lambda+\beta)-(\lambda-1)(\alpha+\beta)\left(2\beta\left(r^2-1\right)+1\right)-2(\alpha+\lambda)(\beta-\lambda)}.$$

where $\lambda > 0$. Thus, for instance, if $t = 0$ we obtain the largest and smallest singular values of $A$ as $\lambda = 1$ and $\lambda = \beta$. Then swapping $\lambda$ for $-\lambda$ (which does not change (12)) we find the middle singular value at $t = 0$ is $\lambda = \alpha$. More generally, this equation gives us a cubic in $\lambda$ from which we can find closed form solutions for the singular values of $(A + tB_0)^T(A + tB_0)$. From this, we get $\lambda_{max}$ and $\lambda_{min}$ and replacing $\lambda \leftrightarrow -\lambda$ gives $\lambda_{med}$.

## 5 The Singular Values of $(A + tB_0)^T(A + tB_0)$

As an example, we illustrate the case $A \in Sing(1, 3, 11)$.

The singular values of the matrix $(A + tB_0)^T(A + tB_0)$ are shown in Fig. 7. Note the loss of smoothness as the singular values cross. The regular branches of the maximum and minimum eigenvalues are increasing in $t$ and do not cross, while that of the middle eigenvalue is decreasing in $t$. The middle eigenvalue crosses each of the largest and smallest eigenvalues just once at typically asymmetric points ($t_- < 0$ for the mid and max eigenvalues, and $t_+ > 0$ for the mid and min eigenvalues). These crossings typically yield different values of the linear distortion. The limiting jump is determined by the largest of these two values.

For each real number $t$, the three singular values of $(A + tB_0)^T(A + tB_0)$ are given by the three different choices of $\zeta, \bar{\zeta}, -1$ as solutions to $\zeta^3 = -1$ from the formula below. These occur from the classical formula for the solution of a cubic equation.

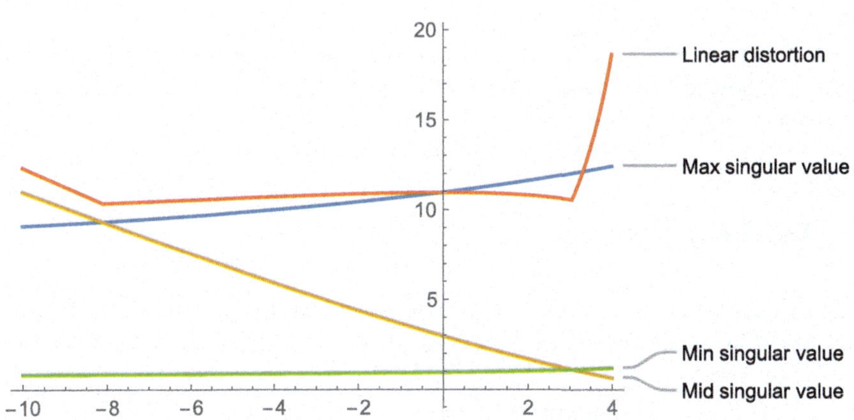


**Fig. 7** The eigenvalues of matrix $(A + tB_0)^T(A + tB_0)$

$$-\frac{C}{3D}-\frac{\varsigma\sqrt[3]{\sqrt{(27\,AD^2-9BCD+2C^3)^2-4\,(C^2-3BD)^3}-27\,AD^2+9BCD-2C^3}}{12\sqrt[3]{2}D}$$
$$+\frac{\bar{\zeta}\,(3BD-C^2)}{6\,2^{2/3}D\sqrt[3]{\sqrt{(27\,AD^2-9BCD+2C^3)^2-4\,(C^2-3BD)^3}-27\,AD^2+9BCD-2C^3}}.$$

Here, in terms of the singular values of $A$ we have,

$$\begin{aligned}
A &= 2\alpha\beta(\beta+1)\left(\alpha^2+(\alpha-1)\beta+\alpha+\beta^2+1\right)-\beta t\left(2\alpha^2+2\alpha(\alpha+4)\beta+\beta^3+\beta^2+\beta+1\right)\\
B &= t\left(2\alpha^2(\beta+1)^2+\alpha(\beta(\beta+6)+1)(\beta+1)+\beta(\beta(\beta(\beta+4)-2)+4)+1\right)\\
&\quad -2(\beta+1)\left(\alpha^2+(\alpha-1)\beta+\alpha+\beta^2+1\right)\left(\alpha(\beta+1)-\beta\right)\\
C &= -2(\beta+1)(-\alpha+\beta+1)\left(\alpha^2+(\alpha-1)\beta+\alpha+\beta^2+1\right)-2(\beta+1)t\left(\alpha^2+(\alpha-1)\beta+\alpha+\beta^2+1\right)\\
D &= 2(\beta+1)\left(\alpha^2+(\alpha-1)\beta+\alpha+\beta^2+1\right).
\end{aligned}$$

It seems quite remarkable that the discriminant admits a simple quadratic factor, but this is a consequence of the fact there are only two eigenvalue crossings. This quadratic factor, given below, determines the crossing points of the singular values.

$$\begin{aligned}
0 = {}& 2\Big(\alpha^4\beta^2+2\alpha^4\beta+\alpha^4-2\alpha^2\beta^3-4\alpha^2\beta^2-2\alpha^2\beta-\alpha\beta^5-\alpha\beta^4+2\alpha\beta^3+2\alpha\beta^2-\alpha\beta\\
&-\alpha+\beta^5+\beta^4+\beta^2+\beta\Big)\\
&-t_\pm\Big(4\alpha^3\beta^2+8\alpha^3\beta+4\alpha^3+\alpha^2\beta^3+7\alpha^2\beta^2+7\alpha^2\beta+\alpha^2+2\alpha\beta^4-4\alpha\beta^3-12\alpha\beta^2\\
&-4\alpha\beta+2\alpha-\beta^5-6\beta^4-\beta^3-\beta^2-6\beta-1\Big)\\
&-t_\pm^2\Big(-2\alpha^2\beta^2-4\alpha^2\beta-2\alpha^2-\alpha\beta^3-7\alpha\beta^2-7\alpha\beta-\alpha-\beta^4-4\beta^3+2\beta^2-4\beta-1\Big).
\end{aligned}$$

At this point, we have sketched how to find the best direction, and given this direction, how to find the maximal interval on which the linear distortion is smooth and concave.

## 6 The Jump Ratio

Our explicit formulas now allow a computational investigation to help understand the size of the jump. Given $A \in Sing(1,\alpha,\beta)$ we define the jump as the value of

$$\frac{1}{\beta}\min\{\mathcal{H}(A+tB_0) : t_- \le t \le t_+\} \le 1$$

Recall the numbers $\alpha, \beta$ determine both $B_0$ and the values $t_-$ and $t_+$ where the singular values of $A + tB_0$ cross.

These examples suggest that the largest energy deficiency can be as much as a factor of $\sqrt{2}$ - that is the relative jump is $1/\sqrt{2}$ and that occurs for a linear deformation $A \in Sing(1, c, c^2)$ as $c \to \infty$. It is a rather lengthy calculation we do not present here to show that any number less than $\sqrt{2}$ is achievable. However, we cannot yet prove that this is best possible and that $A \in Sing(1, c, c^2)$ is a critical case. However, we do have the following theorem.

**Theorem 4** (Largest Jump) *Let $c < \sqrt{2}$. Then there is a sequence of $H$-quasiconformal mappings converging to an $H_\infty$-quasiconformal mapping (and no smaller) with*

$$H_\infty > c\, H$$

## 7 Limit Regimes—One and Two Dimensions

In this section, we consider various limit regimes modelling the situation in one or two dimensions by letting $\alpha, \beta \to \infty$ in various ways. This has the same effect as letting the smallest singular value (which we have normalised to 1 in our discussion above) tend to 0 and effectively removing a variable from the deformation.

We have outlined above the closed form solutions for the two values where the singular values cross $t_\pm$ and therefore have for $A = Sing(1, \alpha, \beta)$ the following limits.

$$\lim_{\beta\to\infty} t_+ = 2(\alpha - 1)$$
$$\lim_{\beta\to\infty} t_- = -\infty.$$

The angle $\theta$ between the normal to the lamination with the principal direction $(0, 0, 1)$ is found using our formula for $u$ above. It is

$$\cos(\theta) = \frac{(\beta - 1)\sqrt{\beta}}{\sqrt{2}\sqrt{(\beta + 1)\left(\alpha^2 + (\alpha - 1)\beta + \alpha + \beta^2 + 1\right)}}$$

.

We say a deformation is strongly anisotropic if $A \in Sing(1, \alpha, \beta)$, $\beta >> 1$ and $\beta/\alpha$ is large. Examples are $A \in Sing(1, c, c^2)$ if $c$ is large. Such examples are illustrated above in Figs. 7 and 8.

For a strongly anisotropic deformation, we find the following "limiting angle" between the lamination and the principle direction.

**Quasi 1-Dimensional Case:** ***or strongly anisotropic deformation***:

$$\alpha/\beta \to 0 \text{ we get the angle } \frac{\pi}{4}.$$

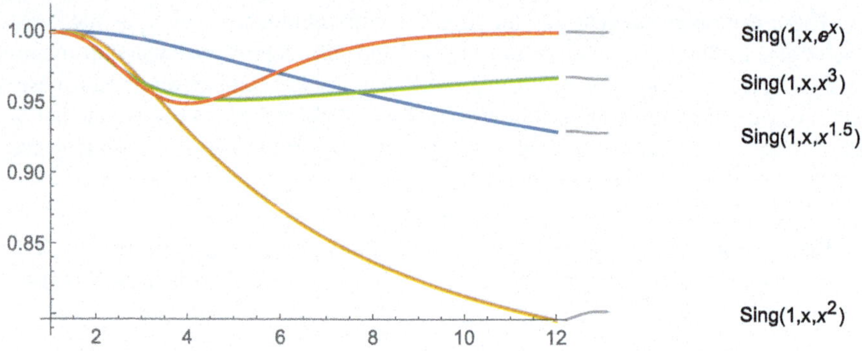


**Fig. 8** The relative jump for $A \in Sing(1, x, f(x))$ where $f$ is indicated by the labelling

There appears some evidence to support this value in Ge et al. (2004). Also examples can be seen in Ball and Schryvers (2003, Fig. 1) of crossing type macrotwins at angle $\frac{\pi}{4}$.

**Quasi 2-Dimensional Case:** ***or weakly anisotropic deformation***:
If $\beta = k\alpha$, $0 \le k < 1$, then

$$\cos(\theta) = \frac{1}{\sqrt{2}\sqrt{k^2 + k + 1}}.$$

Thus, if $\beta - \alpha$ is bounded and $\beta \to \infty$, then $\beta/\alpha \to 1$ and the limiting angle is $\cos^{-1}(\frac{1}{\sqrt{6}})$. We illustrate examples of the relative jumps in these cases. Note the lack

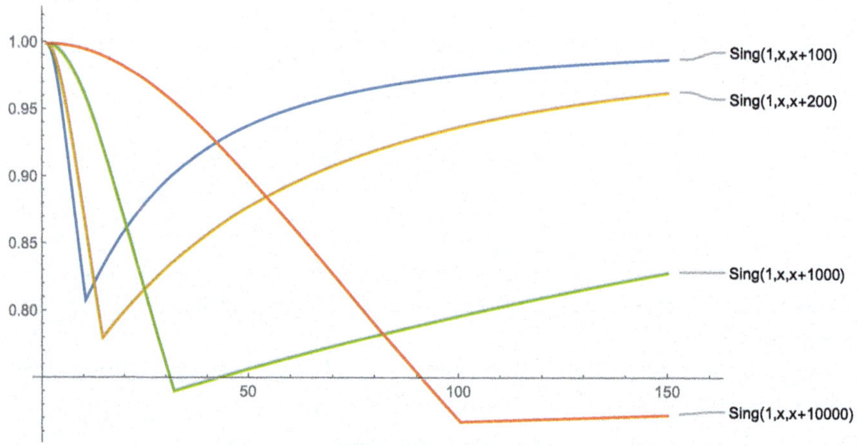


**Fig. 9** The relative jump for $A \in Sing(1, x, f(x))$ where $f$ is indicated by the labelling. Here $x - f(x)$ bounded as $x \to \infty$

of smoothness, this appears to be caused as a switch between the values $\mathcal{H}(A + t_- B_0)$ and $\mathcal{H}(A + t_+ B_0)$ while both lie below the value $\beta$. It is to be remarked on, in light of the suggestions above about $A \in Sing(1, c, c^2)$, that this loss of smoothness appears to be near $\sqrt{\beta - \alpha}$ (Fig. 9).

## References


Astala K, Iwaniec T, Martin GJ (2009) Elliptic partial differential equations and quasiconformal mappings in the plane, Princeton Mathematical Series, 48. Princeton University Press, Princeton, NJ

Astala K, Iwaniec T, Martin GJ, Onninen J (2005) Extremal mappings of finite distortion. Proc Lond Math Soc 91:655–702

Ball J, Défauts dans les cristaux et dans les cristaux liquides, lecture notes. https://people.maths.ox.ac.uk/ball/Teaching/parisox.pdf

Ball J (1981) Global invertibility of Sobolev mappings and interpenetration of matter. Proc R Soc Edinburgh 88:315–328

Ball J, Carstensen C (1999) Compatibility conditions for microstructures and the austenite-martensite transition. Mater Sci Eng A 273–275:231–236

Ball JM, James RD (1987) Fine phase mixtures as minimizers of energy. Arch Rational Mech Anal 100:13–52

Ball JM, Schryvers D (2003) The analysis of Macrotwings in NiAl martensite. J Phys IV France 112:52–159

Bhattacharya K (2003) Microstructure of martensite; Why it forms and how it gives rise to the shape-memory effect. Oxford series on materials modelling. Oxford University Press, Oxford

Dacorogna B (2009) Introduction to the calculus of variations, Translated from the 1992 French original, 2nd edn. Imperial College Press, London

Ge Y, Heczko O, Söderberg O, Lindroos VK (2004) Various magnetic domain structures in a Ni-Mn-Ga martensite exhibiting magnetic shape memory effect. J Appl Phys 96:21–59

Hencl S, Koskela P (2006) Regularity of the inverse of a planar Sobolev homeomorphism. Arch Ration Mech Anal 180:75–95

Heinonen J, Koskela P (1995) Definitions of quasiconformality. Invent Math 120:61–79

Hencl S, Koskela P (2014) Lectures on mappings of finite distortion, Lecture notes in mathematics, vol 2096, Springer, Berlin

Iwaniec T (1998) The failure of lower semicontinuity for the linear dilatation. Bull Lond Math Soc 30:55–61

Iwaniec T, Kovalev LV, Onninen J (2013) Lipschitz regularity for inner variational equations. Duke Math J 162:643–672

Iwaniec T, Martin G (2001) Geometric function theory and non-linear analysis. Oxford Mathematical Monographs. The Clarendon Press, Oxford University Press, New York

Iwaniec T, Martin GJ, Onninen J (2014) On minimisers of $L^p$-mean distortion. In: Computational methods and function theory, vol 14, pp 399–416

Iwaniec T, Onninen J (2016) Smoothing defected welds and hairline cracks. SIAM J Math Anal 48:281–301

Iwaniec T, Onninen J (2010) Neohookean deformations of annuli, existence, uniqueness and radial symmetry. Math Ann 348:35–55

Martin GJ (2015) The tension equation with holomorphic coefficients, harmonic mappings and rigidity. Complex Var Elliptic Equ 60:1159–1167

Martin GJ, McKubre-Jordens M (2012) Minimizers of mean distortion and the Grötzsch problem. J Lond Math Soc 85:282–300
Martin GJ, Yao C (2020) The $L^p$-Teichmüller Theory; existence and regularity. To appear
Martin GJ, Yao C (2021) Non-variational extrema of exponential Teichmüller spaces. Math. Arxiv:1910.13079

# Water Usage Monitoring System for Plants Based on a Weight Sensor and Imagery Motion Analysis

**Astriati Hamidah, Takashi Okayasu, Yu Hoshino, Eiji Inoue, Yasumaru Hirai, Muneshi Mitsuoka, Andri Prima Nugroho, and Lilik Sutiarso**

## 1 Introduction

Monitoring plant growth status, water usage and the surrounding environmental conditions is important for improving plant cultivation techniques. By understanding plant response due to water stress and environmental influence, it is possible to manage water usage based on plant requirements, such as the implementation of efficient irrigation strategies (Dhillon et al. 2014). Information on water usage and requirements can be obtained through several methods, that is, the use of soil sensors, estimating evapotranspiration and water balance measurements. For certain individual observations, the water balance method is reliable for specifying the water flow of a plant system. Therefore, it is a good approach to determine the timing and amount of irrigation requirements (Campos et al. 2016).

By adopting the water balance approach, a weight-based lysimetric method can be used to determine the elements of water balance (Jiménez-Carvajal et al. 2017). The advantage of the weighing lysimeter is that it minimises damage to the target plants (Beeson 2011). In contrast, plant body behaviour is significantly influenced by its water content. One approach of determining the behaviour is realised by plant body

A. Hamidah
Kyushu University Graduate School of Bioresource and Bioenvironmental Sciences, 744 Motooka, Nishi-ku, Fukuoka, Japan

T. Okayasu (✉) · E. Inoue · Y. Hirai
Kyushu University Faculty of Agriculture, Fukuoka, Japan
e-mail: okayasu@bpes.kyushu-u.ac.jp
URL: https://www-ampsd.bpes.kyushu-u.ac.jp/

Y. Hoshino
Faculty of Egineering, Kyushu University, Fukuoka, Japan

M. Mitsuoka
Faculty of Agriculture, University of the Ryukyus, Nishihara, Japan

A. P. Nugroho · L. Sutiarso
Department of Agricultural and Biosystems Engineering, Universitas Gadjah Mada, Yogyakarta, Indonesia



R. McKibbin et al. (eds.), *Proceedings of the Forum "Math-for-Industry" 2019*, Mathematics for Industry 36, https://doi.org/10.1007/978-981-19-1154-5_5

motion analysis, which mainly focuses on plant movement as the physical indicator of plant water stress (Nugroho et al. 2016).

In this study, a monitoring system for plant weight, environmental conditions and plant body motion behaviour using low-cost devices was developed and its accuracy was verified by a cultivation experiment in a greenhouse.

# 2 Materials and Methods

## *2.1 Overview of the Measurement Setup*

The study was conducted in a greenhouse at a university farm in the Faculty of Agriculture, Kyushu University. Tomato plants (Takii Middy 165, *Solanum lycopersicum L.*) were cultivated in the greenhouse. An individual tomato plant pot was arranged in 50 × 50 × 200 cm chamber for measurement of the plant weight and growth. Irrigation water with dissolved chemical fertiliser was applied every hour at a constant amount. The overall measurement system is depicted in Fig. 1.

## *2.2 Arduino-Based Measurement System for Plant Weight and Environmental Condition*

The Arduino-based measurement system was used to measure environmental conditions and plant weight. The environmental data (ambient air temperature and humidity) was recorded using a SHT-35 sensor. The load cell sensor (LCB03, A&D), with 10 kg capacity was selected for the weight measurement. A temperature sensibil-

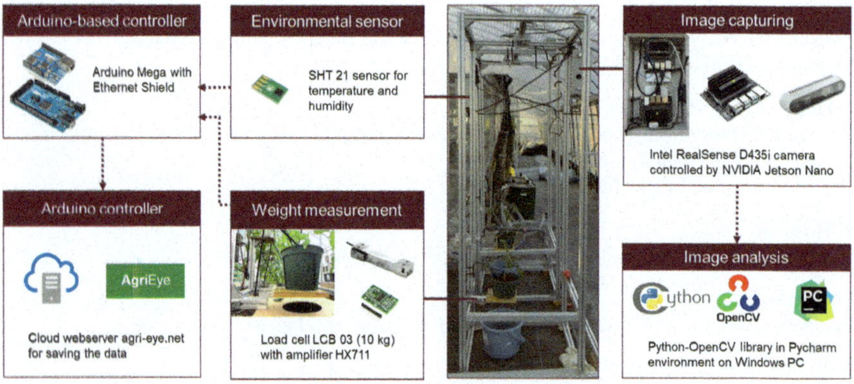


**Fig. 1** Experimental setup for the measurement system in the greenhouse

ity test was conducted at temperature range of 5–40 °C. The LCB03 showed small weight fluctuations within ±3 g tolerance for all investigated temperature ranges. To measure target plant weight efficiently, two load cell units were used. One load cell was installed at the bottom of the cultivation pot and another load cell was installed at the top of the chamber with a hanging wire as shown in Fig. 1. As tomato plants grow bigger, the inclination caused by shoot growth is expected. Therefore, the second load cell was installed at the upper part of the plant to support the plant and the weight was measured by the loaded cell with the hanging wire.

## 2.3 Plant Image Acquisition and Motion Analysis Using the Optical Flow Method

Plant images were captured using an RGB-D camera (RealSense D435i, Intel), which was controlled by a Jetson Nano microcomputer (NVIDIA). Jetson Nano is a small and powerful computer for embedded applications, artificial intelligence (AI) and IoT. The camera can capture both RGB images and depth images obtained from the active stereoscopic technology using an infrared camera with random-dot pattern projection. The plant body motion was calculated from the moving distance of the feature points between two consecutive image frames at time $t = t$ and time $t = t + \Delta t$ as shown in Fig. 2. The feature points were chosen using the Shi-Tomasi corner detector based on the given criteria. These criteria are used to obtain the tracking points, thus the values were carefully chosen to obtain significant points and ignore background noise. The Pyramid Lukas-Kanade criteria are important for obtaining the feature of optical flow. The points from two frames were tracked using the optical flow function. The pixel from the first frame is considered as $I(x, y, t)$, and displacement vectors between two different image frames are described by the distance $\Delta x$ and $\Delta y$. In this computation, we assumed that the brightness between the images does not change, so that the position of each pixel in the next frame becomes $I(x, y, t) = I(x + \Delta x, y + \Delta y, t + \Delta t)$. The displacement distance of the two vectors was computed.

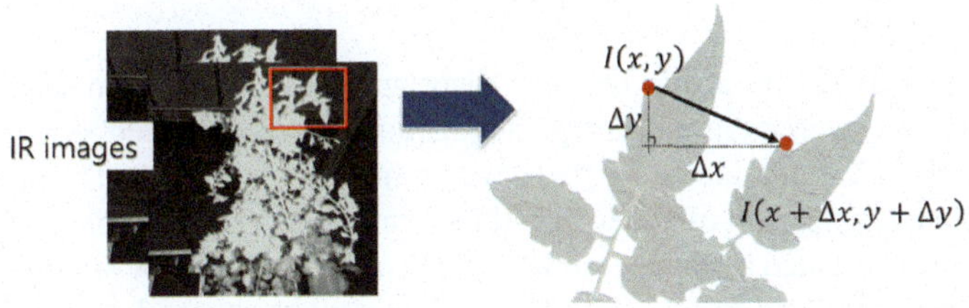


**Fig. 2** Point tracking between two consecutive frames using optical flow method

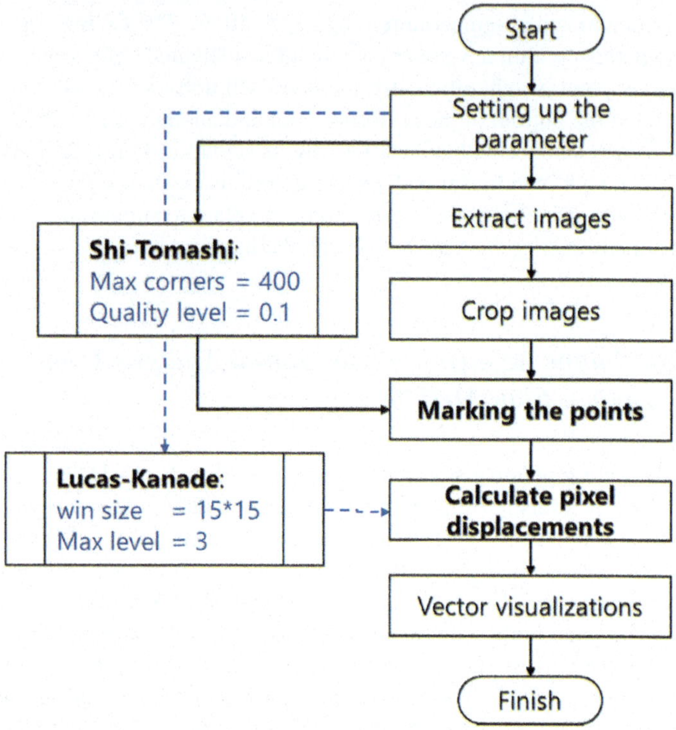


**Fig. 3** Algorithm of Python script for optical flow analysis

Plant body motion analysis using the optical flow method was established using the OpenCV image library (Intel) and a Python script. The algorithm for the script is illustrated in Fig. 3.

## 2.4 *Experiment of Weight Measurement and Image Acquisition*

To understand the relationship between plant water usage and plant body motion, plant weight measurement and image acquisition were conducted simultaneously. Data collection was conducted for five days, from 5 to 9 July 2020. The plant weight was calculated as the sum of two load cell measurements. The dynamics of the water balance were estimated by the measured weight, as the soil and pot weights remained constant, while the soil water content and plant weight changed dynamically. The amount of water absorbed by the target plant roots was considered as evapotranspiration, since the water was simultaneously lost due to the process of evaporation from the soil surface and transpiration through the stomata. In addition, part of the water

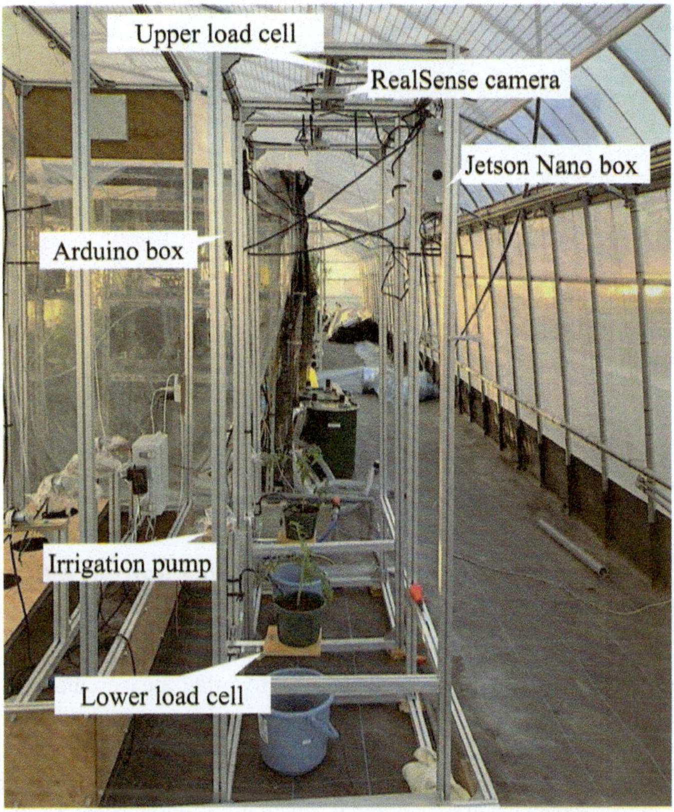


**Fig. 4** Setup of the weight measurement and image acquisition experiment

was used for the plant photosynthesis. Plant images were captured by a RealSense camera attached to the top of the chamber. The setup for weight measurement and image acquisition is depicted in Fig. 4.

# 3 Results and Discussion

## *3.1 Determination of Plant Body Motion Using Optical Flow Analysis*

Plant body motion analysis was conducted to estimate the wilting behaviour induced by the loss of water balance in the plant. Infrared (IR) images were used to measure the plant body motion, since the difference in the IR image quality between day and nighttime was smaller than that of the RGB images. Figure 5 shows the detection

**Fig. 5** Plant body motion detection procedure

procedure for plant body motion using the optical flow method. The captured IR images were provided as input images to compute the optical flow, and cropped images with plant body motion were subsequently obtained. Plant body motion was visualised using several arrows representing the distance and direction corresponding to the motion of the target plant body.

### 3.2 *Effect of Environmental Conditions on Plant Weight and Motion Displacement*

Figure 6 shows time-series data obtained from five-day measurement results (5–9 July 2020). Graphs A and B show ambient air temperature and humidity. Graph C shows the plant weight variation directly measured by the two load cells. Graph D shows the displacement of the target plant body motion computed using the optical flow method. As seen in the graphs A and B, fluctuations of temperature and humidity during daytime and nighttime became small in the rainy days (Day 2 and 3), while they were larger on the sunny day (Day 1 and 4). From the graph C, the variation of plant weight during daytime and nighttime in the rainy days became smaller than the values measured during sunny days. This tendency was also observed in the plant body motion data (graph D).

Figure 7 shows a comparison between the hourly weight variation and displacement on a rainy day (Day 3). The data was computed from the fluctuation of the weight measured every hour, and each displacement data point was obtained by the mean value of the displacement of the plant body motions. The weight variation on the rainy day (Day 3) was small, since the water quantity absorbed from the roots was limited due to the low evapotranspiration rate (low rate of photosynthesis). Consequently, water provided by irrigation was immediately discharged from the bottom of the pot. Furthermore, the hourly displacement of the plant body was also remarkably small (the maximum mean displacement was 2 pixels), as shown in Fig. 8. The plant did not show the movement; that is, wilting behaviour did not occur during the rainy days.

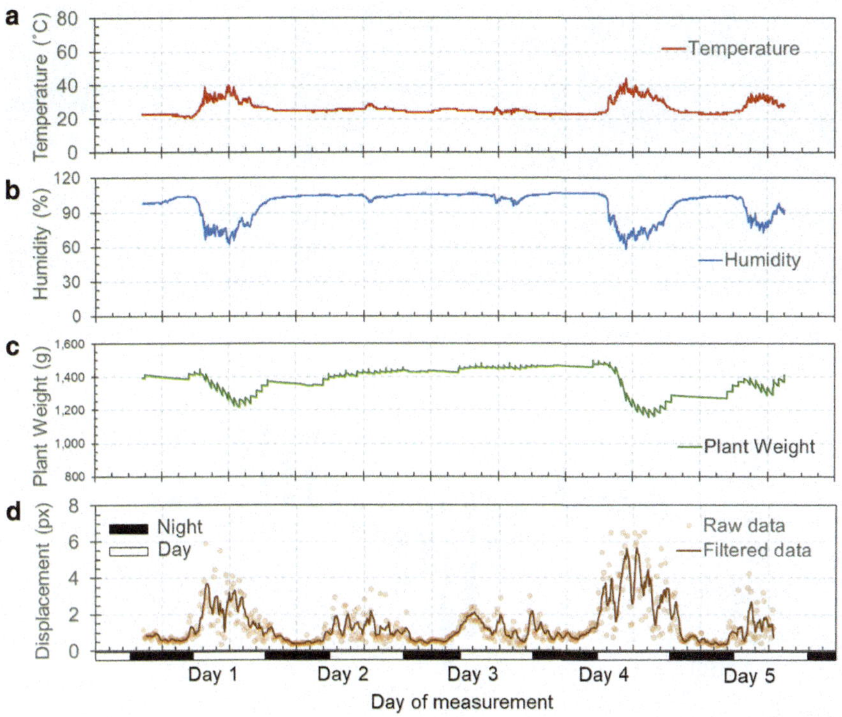


**Fig. 6** Time series data of temperature (**a**), humidity (**b**), pot weight (**c**), and plant body motion displacement (**d**) over five days measurement

**Fig. 7** Hourly weight variation and displacement of rainy day condition (Day 3)

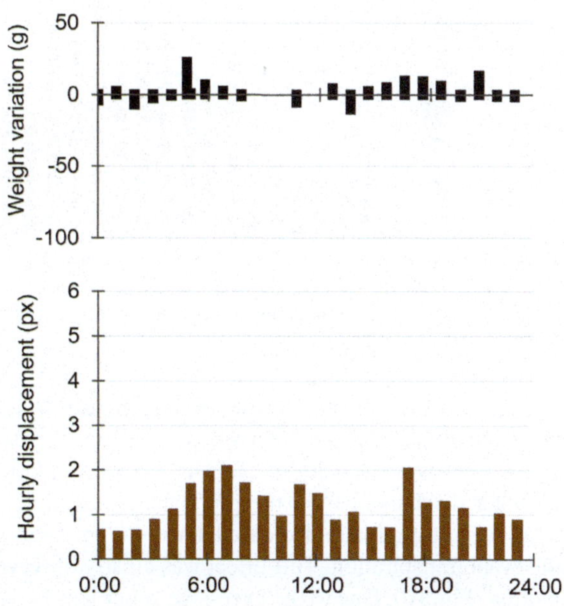

**Fig. 8** Visualisation of wilting behaviour in the morning (09:00), afternoon (13:00) and evening (18.00) on a rainy day (Day 3)

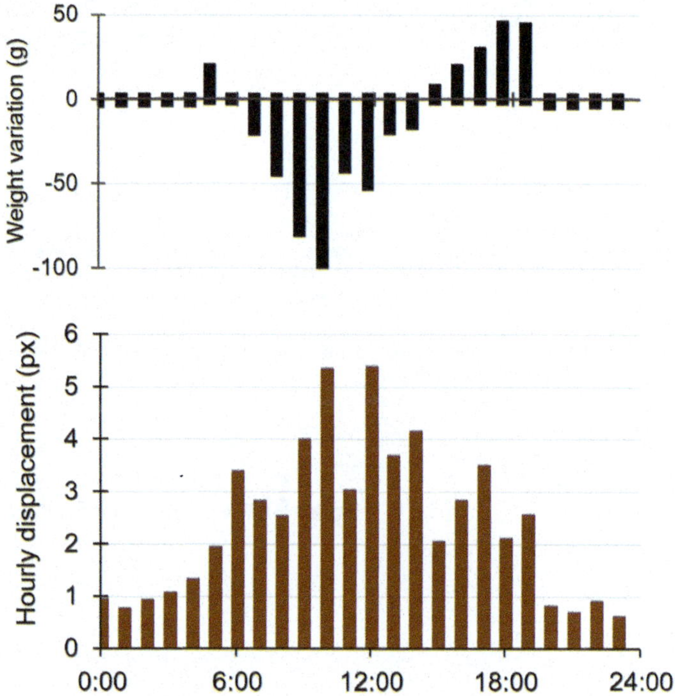


**Fig. 9** Hourly weight variation and displacement of sunny day condition (Day 4)

Figure 9 shows a comparison between hourly weight variation and displacement on a sunny day (Day 4). Plant weight decreased sharply from the morning because of the improved capacity of the soil to absorb and hold water so that the amount of water discharged from the pot became insignificant. However, in this test, the plant weight gradually decreased since the provided water was not enough against the evapotranspiration. This effect was clearly observed from the hourly plant body motion (Fig. 10). Due to the excessive water loss from the plant body by high evap-

**Fig. 10** Visualisation of wilting behaviour in the morning (09:00), afternoon (13:00) and evening (18.00) on a sunny day (Day 4)

otranspiration, wilting behaviour of the plant body was evident. The change in the displacement could be used to estimate the water stress of a plant. Moreover, in such water management conditions, plant weight and shape cannot revert back to characteristics of the previous day. It is suggested that the action of excessive water stress is a factor that can cause significant damage to plants. Therefore, the aforementioned findings are likely to become a key factor in building an irrigation management system based on plant behaviour.

## 4 Conclusions

In this study, a measurement system for water usage of a single plant based on a weight sensor technique and image analysis was developed using low-cost devices. The validity of the measurement system was verified via an evaluation test in a greenhouse and the following results were obtained.

1. The weight measurement accuracy was $\pm 3$ g in the prescribed temperature range (5–40 °C). This accuracy was sufficiently high for plant weight measurements.
2. Plant body motion was continuously and effectively measured using time-lapse IR images and the optical flow method.
3. The motion of the single tomato plant was significantly correlated with the weight change. In particular, plant body motion on sunny days was visible, and wilting behaviour was well captured from the motion analyses.

Overall, plant weight variation and plant motion were effectively measured using sensor and image analysis. However, the optimal amount of water for the tomato plant could not be estimated directly from their relation of the above-mentioned variables. This is because it depends strongly on the plant growth stage and cultivation method. The correlation between plant weight and motion will be the object of future studies.

**Acknowledgements** This research was supported by grants from the Japan Society for the Promotion of Science KAKENHI (JP18K05906) and the Japan Science and Technology Agency (JPMJST1714).

## References

Beeson RC (2011) Weighing lysimeter systems for quantifying water use and studies of controlled water stress for crops grown in low bulk density substrates. Agric Water Manag 98(6):967–976

Campos I, Balbontín C, González-Piqueras J, González-Dugo MP, Neale CMU, Calera A (2016) Combining a water balance model with evapotranspiration measurements to estimate total available soil water in irrigated and rainfed vineyards. Agric Water Manag 165:141–152

Dhillon R, Udompetaikul V, Rojo F, Roach J, Upadhyaya S, Slaughter D, Lampinen B, Shackel K (2014) Detection of plant water stress using leaf temperature and microclimatic measurements in almond, walnut, and grape crops. Trans ASABE 57(1):297–304

Jiménez-Carvajal C, Ruiz-Peñalver L, Vera-Repullo JA, Jiménez-Buendía M, Antolino-Merino A, Molina-Martínez JM (2017) Weighing lysimetric system for the determination of the water balance during irrigation in potted plants. Agric Water Manag 183:78–85

Nugroho AP, Okayasu T, Sakai A, Inoue E, Hirai Y, Mitsuoka M, Sutiarso L (2016) Automatic leaf motion analysis using optical flow to diagnose plant behaviour in response to environmental changes. In: Proceedings of 8th ISMAB, vol (5), pp 848–853

# Development of Plant Phenotyping Platform Using Low-Cost IoT Devices and Its Performance Evaluation

**Jiro Ito, Takashi Okayasu, Koichi Nomura, Daisuke Yasutake, Tadasige Iwao, Yukio Ozaki, Eiji Inoue, Yasumaru Hirai, and Muneshi Mitsuoka**

## 1 Introduction

Agriculture in Japan faces various challenges such as the ageing of farmers, frequent abnormal weather, and the diversity of consumers' preferences and demands. The number of people working in the agricultural sector decreased from 2.606 million in 2010 to 1.686 million in 2019, a decrease of 35% (Ministry of Agriculture 2019). Moreover, new farmers are expected to play a role in the future of Japanese agriculture. However, approximately 30% of them leave within five years owing to unstable living (Ministry of Agriculture 2014), suggesting that know-how is not transferred successfully. The Japanese government actively promotes research and development to reconstruct our food supply system from agricultural production to distribution and consumption. These have been deployed in precise field environments and through plant growth measurements based on the Internet of Things (IoT) and artificial intelligence, their evaluation and analyses and the dissemination of the results.

---

J. Ito
Graduate School of Bioresource and Bioenvironmental Sciences, Kyushu University, West Bldg. 5 877, 744 Motooka, Nishi-ku, Fukuoka, Japan

T. Okayasu (✉) · D. Yasutake · Y. Ozaki · E. Inoue · Y. Hirai
Faculty of Agriculture, Kyushu University, Fukuoka, Japan
e-mail: okayasu@bpes.kyushu-u.ac.jp
URL: https://www-ampsd.bpes.kyushu-u.ac.jp/

K. Nomura
Faculty of Agriculture and Marine Sciences, Kochi University, Kochi, Japan

T. Iwao
Fujitsu Limited, Tokyo, Japan

M. Mitsuoka
Faculty of Agriculture, University of the Ryukyus, Nishihara, Japan

 
R. McKibbin et al. (eds.), *Proceedings of the Forum "Math-for-Industry" 2019*, Mathematics for Industry 36, https://doi.org/10.1007/978-981-19-1154-5_6

In particular, precise measurement of plant growth and development under various environmental conditions is indispensable for advancing plant breeding and cultivation management technology. An example is a large phenotyping platform in the form of a gantry crane (Nicolas et al. 2016), which facilitates many phenotypic analyses. Plant height is a critical factor in measuring plant growth and is determined by RGB colour and Depth information (RGB-D) cameras (Kurosaki et al. 2017). However, many of these techniques are expensive, and the development of low-cost techniques is slow.

In this study, we employed affordable IoT devices, and open-source software and hardware to develop a plant phenotyping platform that can automatically measure microclimatic environmental data and plant growth and development information in the plant growth space. In addition, the developed platform was used to automatically measure the growth of spinach cultivated hydroponically in a horticultural house, and the performance and the measurement accuracy of the platform were evaluated.

## 2 Materials and Methods

### *2.1 Overview of the Phenotyping Platform*

Figure 1 shows the overall structure of the developed phenotyping platform. The platform consisted of a phenotyping robot with a four-wheel travelling unit and a two-wheel motor drive system, and an aluminium rail unit setup on both sides of the hydroponics grow bed. A single-board computer called Raspberry Pi 3B (raspberrypi.org) was mainly used for the robot control operation. This platform provided image-recognition-based location detection and control functions using AR markers attached to the rail unit. The camera used for marker recognition had a near-infrared light-emitting diode (LED) module to capture photographs both in the day and at the night.

During the actual measurements, recognition errors and failures might occur owing to human error and environmental factors induced by natural light. However, the accuracy of position control was less than 3 mm in both daytime and nighttime (Ito et al. 2021). Limit switches for stopping the motor power supply were placed at the front and rear parts of the robot to prevent loss of control and several accidental operational failures.

### *2.2 Plant Cultivation and Imaging Test*

The measurement test period was from 30 December 2019 to 10 February 2020. The measurements were performed at 3:00, 8:00, 11:00, 15:00, 19:00 and 23:00 each day at seven locations. Spinach (Spinacia oleracea) was used as the experimental

**Fig. 1** Structure of the phenotyping platform

plant and the cultivar was the Wase clone. We applied a nutrient-film technique for spinach cultivation. OAT house fertiliser No. 1 and 2 (OAT Agrio Inc., Tokyo, Japan) were utilised to prepare the culture medium. Fertiliser No. 1 comprises a balanced combination of essential elements of the crop, whereas the fertiliser No. 2 contained nitrogen and calcium as its main elements. The EC (electrical conductivity), pH and nutrient temperature values were adjusted to 2.0 ms/cm, 5.5–6.0 and approximately 17 °C, respectively. Forty-eight spinach plants were transplanted into a growth panel with 48 holes per panel, and we used seven panels for the cultivation tests. The performance of the developed platform was evaluated based on the utilisation rate which is defined as the ratio of the number of successful measurements to the total number of measurements.

## 2.3 *Plant Height Extraction Using RGB-D Sensor*

An RGB-D camera (RealSense D435i, Intel) was used to extract the plant growth features such as the colour and shape of the plants. To minimise depth data errors owing to ambient light and surface characteristics of the target plants, we averaged 10 depth data values obtained from 10 measurements conducted at the same location. Figure 2 shows the top views of the RGB image and height heat map of the plants. A smoothing function was used for the depth heat map using a median filter because the plant height data contained noise. Plant height was directly calculated as the

(a) RGB image (b) Depth heat map

**Fig. 2** Heat map with difference between initial data and height after growth

difference between the depth data at the transplantation day (initial condition) and the current depth data.

Subsequently, the following three methods, i.e. (1) no binarisation, (2) binarisation by hue value and (3) binarisation by brightness were investigated to extract the leaf area. The OpenCV function cvtColor was used to convert the images. For the hue conversion, we used the RGB2HSV function, and the conversion of brightness was achieved using the RGB2GRAY function. The threshold was determined using Otsu's method.

The plant height values obtained using the above methods were compared with available measurement results. The accuracy was verified through comparison with the measured heights of 18 plants.

## 3 Results and Discussion

### *3.1 Performance of Unmanned Automatic Driving System*

The operation rate of the robot was 86.9% throughout the test period (43 days). There were two main reasons for measurement failures owing to environmental and human factors. The environmental factor was the lack of sufficient light to capture the markers. In the morning (7:00 a.m.), the sky was bright before sunrise, so the near-infrared light that should have been lit in low light did not turn on, and the failure occurred because the light level did not reach the level needed to recognise the marker. The human factor was a simple mistake attributed to a lack of attention. A simple syntax error occurred when the measurement program was improved during the measurement period. Because of grammatical error, the program did not work properly. In addition, as the system was not designed to notify us of such errors, we could observe the cause of the problem delaying our response.

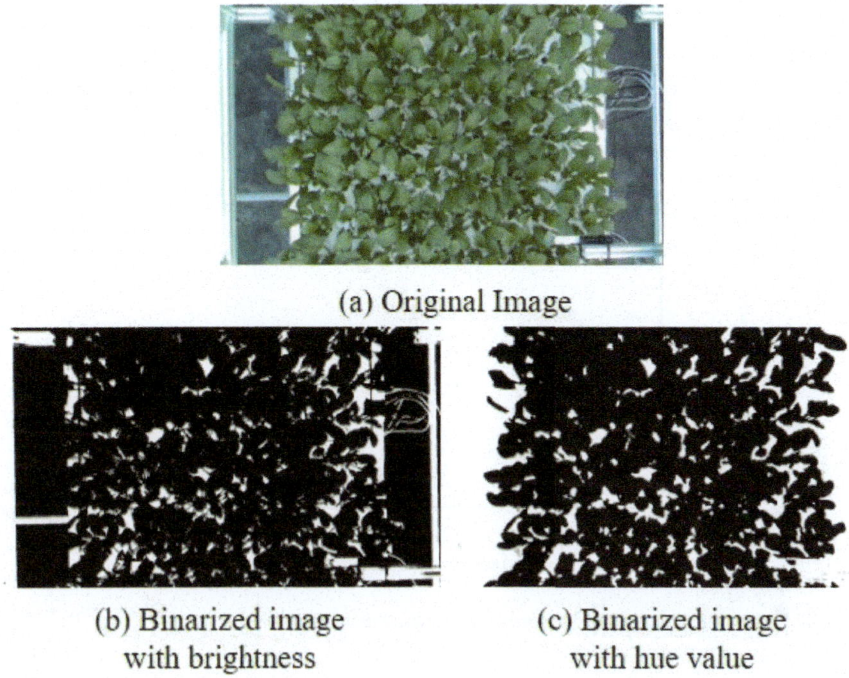


**Fig. 3** Results of the binarisation process

### *3.2 Influence of Different Binarisation Methods on Leaf Region Extraction*

The binarisation results are shown in Fig. 3. The hue value in method (2) was selected as $50 \leq Hue \leq 100$ from the focusing leaf region. The no-binarisation method could not extract the leaf region satisfactorily because of many outliers. The binarisation by brightness (b) could accurately extract the regions of the plant bodies and frames. However, method (b) failed to extract the tips of a leaf owing to the light reflection of the leaf surface when the growth approached the harvesting stage. In addition, this method eliminates the rail frame for the phenotyping robot. However, the binarisation method using hue value (c) could extract the region of plants more effectively than methods (a) and (b).

### *3.3 Accuracy of Plant Height Estimation*

Figure 4 shows a comparison between the estimated and measured plant height values at the mid-growth and harvesting stages. The distribution of the plant was estimated

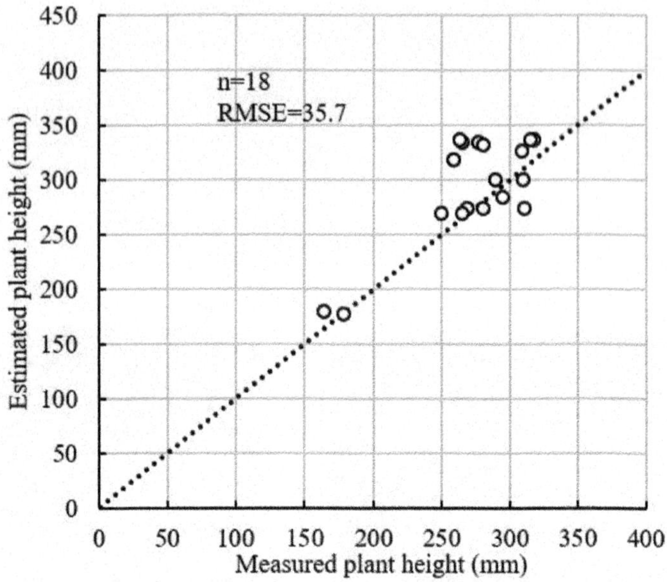


**Fig. 4** Comparison between estimated and measured plant heights

using a linear equation ($y = x$). However, as the plant height increased, the leaves grew to overlap with neighbouring plants, making it difficult to determine the exact plant height. In this experiment, we surveyed the plant heights from the middle to late stages of growth, which caused this problem. However, in the early stage of growth, the leaves did not overlap, so it was possible to obtain more accurate values than in the middle and late stages. In the future, it will be necessary to conduct measurements at detailed growth stages, from planting to harvesting, to investigate the accuracy of the measurements at each growth stage.

## *3.4 Growth Curve and Spatial Distribution of Plants*

Figure 5 shows time variation of the plant height and its spatial distribution. The bold black line represents the mean value of the plant heights. The growth curve of each plant could be traced well using the RGB-D images, and the unevenness of the plant growth could be grasped to some extent. A decrease in the plant height observed at the harvesting stage could be explained as the estimation error of the RGB-D camera owing to the plant bending. We observed that our proposed estimation method could not distinguish whether such estimation error was induced by the bending of leaves or the poor growth.

However, several large outliers were confirmed on 4 January. After checking the cause against the actual image, it was found that the colour of the images was

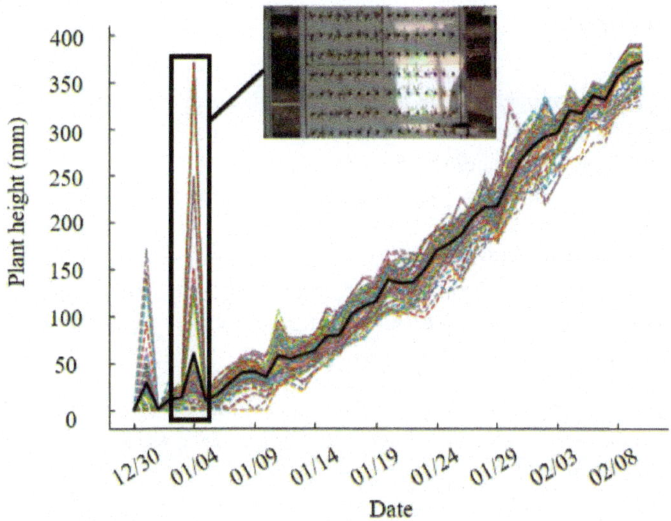


**Fig. 5** Daily variations in plant height

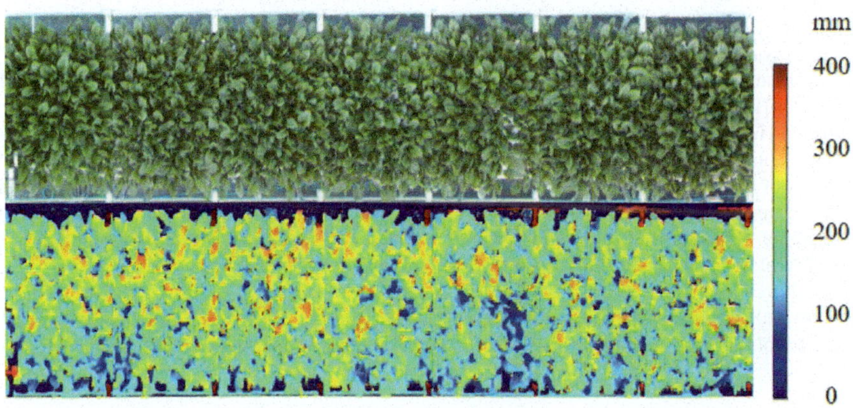


**Fig. 6** Plant height distribution across hydroponics grow bed

partially saturated owing to the effect of strong sunlight. This problem affected the accuracy of the depth measurements because the RGB-D camera utilised the active IR technology to measure the depth information.

Figure 6 shows the plant height distribution across the entire hydroponic grow bed. This type of image was helpful for the quantitative evaluation of growth features in the regions of interests (ROIs) such as leaf colour and plant height. Using an in-house manufactured robot, we could rapidly estimate the growth features of the plants from a broad perspective.

## 4 Conclusions

In this study, we developed a plant phenotyping platform using low-cost IoT devices and evaluated its performance through plant growth measurements. We developed a method to calculate the standing height of plants using RGB-D images and several image-processing techniques. We evaluated our platform over time and performed space measurements of spinach plants. The results for spinach using the developed platform could be utilised for the quantitative evaluation of their growth and spatial distribution. Therefore, the proposed platform is a suitable plant phenotyping system for leafy greens that can achieve spatio-temporal measurements of morphological data and has significant cost advantages.

**Acknowledgements** This study was conducted as a part of the Japan Society for the Promotion of Science, KAKENHI Grant Number JP18K05906 and the joint research project between Fujitsu Limited and Kyushu University.

## References

Ito J, Okayasu T, Nomura K, Yasutake D, Iwao T, Ozaki Y, Inoue E, Hirai Y, Mitsuoka M (2021) Development and performance evaluation of a plant phenotyping platform using low-cost IoT devices [In Japanese]. Agric Inf Res 30(2):13–23

Kurosaki H, Umeda D, Iwasaki Y (2017) Measurement of plant height and growth diagnosing for paprika using Kinect sensor [In Japanese]. Plant Environ Eng 29:130–138

Ministry of Agriculture, Forestry and Fisheries of Japan (2019) The agricultural labor force statistics [translated from Japanese.]. https://www.maff.go.jp/j/tokei/sihyo/data/08.html

Ministry of Agriculture, Forestry and Fisheries of Japan (2014) Annual report on food, agriculture and rural areas in Japan FY2014 [In Japanese]. https://www.maff.go.jp/j/wpaper/w_maff/h26/h26_h/trend/part1/chap2/c2_1_03.html

Nicolas V, Sabermanesh K, Sadeghi-Tehran P, Hawkesford MJ (2016) Field Scanalyzer: an automated robotic field phenotyping platform for detailed crop monitoring. Funct Plant Biol 44:143–153

# Why Do Meals During the Resting Time Cause Fat Accumulation in Mammals?—Mathematical Modeling of Circadian Control on Glucose Metabolism

**Akane Hara and Akiko Satake**

## 1 Introduction

The circadian clock of animals on the earth allows the organisms to synchronize their metabolic, physiological, and behavioral activities with the day and night cycle. The central clock located in the suprachiasmatic nucleus (SCN) regulates behavioral rhythms of sleep/wake and fasting/feeding cycles in response to cyclic environmental changes. However, the energy status of cells needs to be rather homeostatic to supply energy constantly under fluctuating environments.

The energy homeostasis is maintained by the orchestration of metabolic pathways stimulated by energy intake and expenditure (Inoki et al. 2012; Kumar Jha et al. 2015). During the active period, it ensures the uptake and storage of energy substrates from food intakes, which is necessary to maintain a high metabolic rate. During the resting period, stored substrates such as glycogen and fat are metabolized to maintain basal energy expenditure. Alterations in the timing of meals may cause desynchronization in the metabolic rhythms and eventually disturb energy homeostasis (Wang et al. 2011). Thus, the synchronization of feeding behaviors and metabolism is critical to the maintainance of internal homeostasis.

With the advance of civilization, increasingly people work in phases that are desynchronized with their biological clock activities. The misalignment of biological and social time could disrupt the clock and energy homeostasis. Previous studies suggested that shift workers have an increased risk of metabolic diseases (Wang et al. 2011; Vetter et al. 2018; Shan et al. 2018).

A. Satake
Kyushu University, Motooka 744, Nishi-ku, Fukuoka 819-0395, Japan

*Present Address:*
A. Hara (✉)
University of Toyama, 2630 Sugitani, Toyama 930-0194, Japan
e-mail: hara.akane32a@gmail.com

 
R. McKibbin et al. (eds.), *Proceedings of the Forum "Math-for-Industry" 2019*, Mathematics for Industry 36, https://doi.org/10.1007/978-981-19-1154-5_7

The liver is mostly responsible for maintaining energy homeostasis through coordinated regulation of metabolic pathways, especially those controlling the synthesis and metabolism of glucose, lipid, cholesterol, and bile acid (Bass and Takahashi 2010). Despite plenty of experimental findings for the involvement of circadian regulation of glucose metabolism in the liver, the optimal regulatory strategy associated with the two different pathways, glycogen and fat production, for the maintenance of energy homeostasis remains poorly understood. A mathematical framework is useful to assess the circadian regulation of the metabolism of glycogen and fat by evaluating the contribution of a time of the day-dependent metabolic processes on the maintenance of energy homeostasis. We developed a mathematical model describing the dynamics of glycogen and fatty acid contents–two major forms of energy storage in the body. Analyses of our model demonstrate that minimization of high serum glucose is realized when glycogen and fatty acid syntheses are peaked in the end and the middle of the active phase, respectively.

## 2 Material and Methods

### 2.1 A Mathematical Model of Glucose Metabolism

To explore the effect of feeding timing on glucose metabolism, we developed a mathematical model that describes the processes of glucose input, consumption, and allocation to glycogen and fat (Fig. 1a). We assume that fat represents fatty acids that are the main energy source available in the human body and other vertebrates. $x_G$, $x_F$, and $x_L$ represent the amount of glucose, fat, and glycogen in a liver, respectively. The fluctuation of the amount of glucose ($x_G$) in a liver is formalized as follows;

$$\dot{x_G} = aL(t) - d_G(t)x_G - (\gamma_L(t)x_G + \gamma_F(t)x_G) + \beta(t)x_L. \tag{1}$$

The amount of glucose increases by food intake $aL(t)$ and decreases by consumption at a fraction of the existing glucose, $d_G(t)$, at time $t$. Glucose also decreases for the use of glycogen and fat synthesis. Glycogen, a polymer of glucose functioning as energy storage, is synthesized at a rate $\gamma_L(t)x_G$. Fatty acids synthesis in the liver is stimulated by glycolysis and lipogenesis that enhance glucose flux for the production of precursors for lipid biosynthesis such as acetyl-CoA. A series of processes for lipid biosynthesis is assumed to occur at a rate $\gamma_F(t)x_G$. The amount of glucose also increases by degradation of glycogen at a rate of $\gamma_\beta(t)$.

We consider two situations for the timing of food intake, (I) synchronized food schedule in which feeding takes place in the active phase and (II) desynchronized food schedule in which feeding takes place in the resting phase (Fig. 1b). We assume that active and resting phases occur from 0 to 12 h and from 12 to 24 h of the day , respectively. Based on this assumption, feeding in the active phase is formalized by representing $L(t)$ in Eq. (1) as;

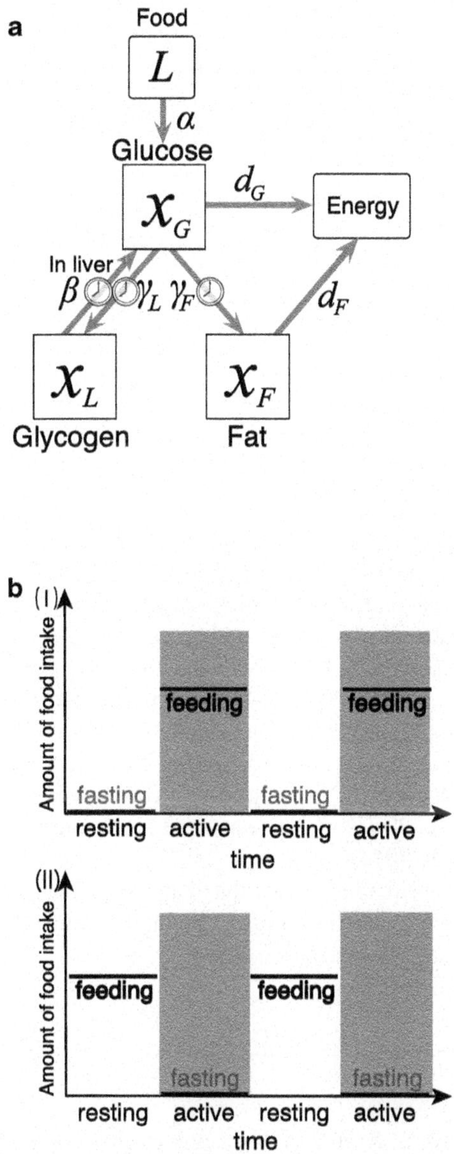


**Fig. 1** **a** *Glucose allocation model.* The amount of glucose, glycogen, and fat are represented by $x_G$, $x_L$, and $x_F$. Glucose is obtained at the rate of $\alpha$ from $L(t)$ food intake depending on the food schedules described in Fig. 1b. Glucose is allocated into glycogen and fat at the rate of $\gamma_L$ and $\gamma_F$, respectively. Glycogen is degraded into glucose at the rate of $\gamma_\beta$. Glucose and fat are consumed as energy resources at the rate of $d_G$ and $d_F$, respectively. $d_G$ is high during the active phase and low during the resting phase (Eq. 3). $\gamma_L$, $\gamma_F$, and $\gamma_\beta$ are regulated by the circadian clock and represented by the cosine curves in Eq. 4. **b** *Two schedules of food intakes.* (I) *Synchronized food schedule.* The active phases and feeding timings are synchronized. We assume that this schedule is natural for mammals, assuming that most mammals obtain food during the active phases. (II) *Desynchronized food schedule.* The active phases and feeding timings are desynchronized. This schedule is assumed to be abnormal habits and similar to the situation of shift workers

$$L(t) = \begin{cases} 1 \; 0 \leq t \bmod 24 \leq 12 \\ 0 \; 12 < t \bmod 24 \leq 24 \end{cases}. \tag{2a}$$

Similarly, feeding in the resting phase is formalized by representing $L(t)$ as;

$$L(t) = \begin{cases} 0 \; 0 \leq t \bmod 24 \leq 12 \\ 1 \; 12 < t \bmod 24 \leq 24 \end{cases}. \tag{2b}$$

We assume that glucose consumption rate is reduced by $\alpha$ in the resting phase as follows:

$$d_G(t) = \begin{cases} d_{GA} \; 0 \leq t \bmod 24 \leq 12 \\ \alpha d_{GA} 12 \leq t \bmod 24 \leq 24 \end{cases}, \tag{3}$$

where $0 < \alpha < 1$.

Three processes in the glucose dynamics in the model are likely to be regulated by the circadian clock. Doi et al. (2010) reported that *Clock* gene mutation in mice damped the circadian oscillation of hepatic glycogen and glycogen synthase 2 (Gys2), suggesting that the rate of glycogenesis indicated as $\gamma_L(t)$ in Eq. (1) is under the influence of the circadian clock. There is also a report showing that a lack of a functional PER2 protein (*Per2$^{Brdm1}$*) resulted in an increased glycogen phosphorylase activity during the fasting phase in mice (Zani et al. 2013). This finding implies that the glycogenolysis rate,$\beta(t)$, is also regulated by the circadian clock. Moreover, the rate of fatty acid production $\gamma_F(t)$ is suggested to be regulated by the circadian clock as the clock-disrupted (*Per1/2 -/-*) mice showed phase shifts of hepatic triglyceride level (Adamovich et al. 2014).

We incorporated the circadian regulation of the rate of fat synthesis ($\gamma_F(t)$), glycogenesis ($\gamma_L(t)$) and glycogenolysis $(\gamma_\beta(t))$ by introducing sinusoidal function with a period of 24 h ($\tau = 24$) for each as follows:

$$\gamma_i(t) = \frac{(a_i - b_i)\cos\left(\frac{2\pi(t-\varphi_i)}{\tau} + 1\right)}{2} + b_i, \tag{4}$$

where $i = F, L,$ and$\beta.a_i$, $b_i$, and $\varphi_i$ correspond to the upper limit, the lower limit, and the phase of each process.

The amount of fat and glycogen, $x_F$, and $x_L$, are formalized as;

$$\dot{x_F} = \gamma_F(t)x_G - d_F(t)x_F, \tag{5}$$

$$\dot{x_L} = \gamma_L(t)x_G - \gamma_\beta(t)x_L. \tag{6}$$

The amount of fat ($x_F$) is increased by the production from glucose at a rate of $b\gamma_F(t)$ and consumed as an energy source at a rate of $d_F$ (Eq. (5)). The amount of glycogen ($x_L$) is increased by the production from glucose at a rate of $\gamma_L(t)$ and decomposed into glucose at a rate of $\gamma_\beta(t)$ (Eq. (6)). Note that $x_F$ and $x_L$ are normalized variables indicating the relative amount of fat and glycogen formed by a single molecule of glucose. We assume that the fat consumption rate $d_F$ is reduced by $\alpha$ in the resting phase as follows:

$$d_F(t) = \begin{cases} d_{GF}\ 0 \le t \bmod 24 \le 12 \\ \alpha d_{GF} 12 \le t \bmod 24 \le 24 \end{cases}. \tag{7}$$

## 2.2 Risks of Energy Homeostasis and Hyperglycemia

We next examined the optimal phase set for glycogen and fat synthesis that minimizes two different risks, energy depletion and hyperglycemia based on the model above. It could be possible to consider that increased fat accumulation by food intake in the resting phase can be the byproduct of minimizing either or both energy homeostasis and hyperglycemia risks in the synchronized feeding schedule in which animals have adopted in nature. According to this assumption, organisms employing the optimal phase set to minimize these risks should reveal increased fat accumulation when they have the desynchronized feeding schedule. We tested this for two different risks by simulations.

The first is the risk of homeostasis in terms of energy availability because it is one of the important aspects of the metabolic system in living organisms. To define energy level at time $t$ ($E(t)$), we here focus on ATPs, the source of energy in many living organisms, which are produced from glucose and fat.

$$E(t) = f_G d_{GL}(t) x_G + f_F d_F(t) x_F. \tag{8a}$$

$f_G$ and $f_F$ indicates the number of ATPs produced from glucose and fat, respectively. From one molecule of glucose, about 30 ATPs are obtained as a result of aerobic respiration and $f_G = 30$. On the other hand, one molecule of palmitic acid, 106 ATPs are produced through beta-oxidation and $f_F = 106$. We consider the relative ATP production efficiency $f = f_F/f_G$ and represent $E^*$ as follows:

$$E^* = E(t)/f_G = d_G(t) x_G + f d_F(t) x_F. \tag{8b}$$

The risk of energy depletion is defined as the cumulative sum of $E(t)$ below the certain threshold as follows:

$$C_E = \int_{n}^{n+48} \{\max\{(E_L - E^*), 0\}\}dt, \tag{9a}$$

where $E_L$ is the lower threshold above which there is no damage of energy depletion. To examine the phase combination that optimizes energy homeostasis, we evaluated $E^*$ for 576 combinations of $\varphi_F$ and $\varphi_L$ by changing each of phase by 1 h.

The second risk is hyperglycemia, which is the major problem of diabetes that can cause neuropathy (Albers and Pop-Busui 2014), retinopathy (Antonetti et al. 2012). The level of hyperglycemia is defined as the cumulative sum of glucose level above a threshold as follows:

$$C_G = \int_{n}^{n+48} \{\max\{(x_G(t) - x_G^U), 0\}\}dt, \tag{9b}$$

where $x_G^U$ is the upper threshold of glucose level below which there is no risk of damage. Examples of the cost of energy exhaustion and high glucose level are presented in Fig. 2a, b, respectively. $C_E$ and $C_G$ are calculated under the condition of synchronized food schedule (Fig. 1b(I)).

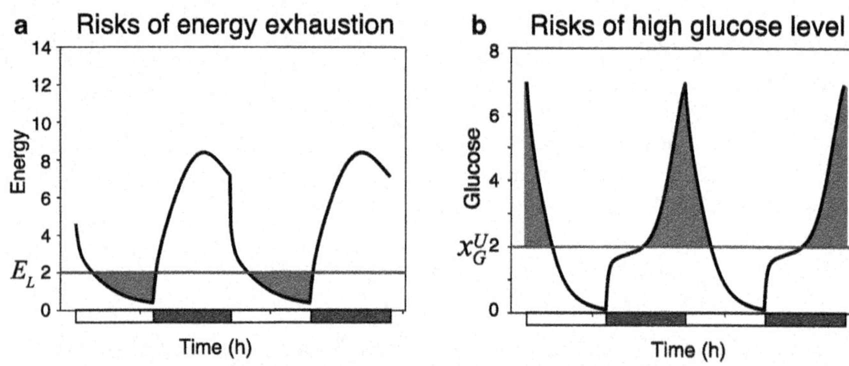


**Fig. 2** *Two types of risks of nutrient homeostasis.* We consider the risks of (**a**) energy exhaustion and (**b**) high glucose level as possible elements behind the nutrient homeostasis. The black and white bars at the time axis represent active and resting phases, respectively. **a** *The risks of energy exhaustion.* The damages of energy exhaustion occur when the $E(t)$ is below the threshold $E_L = 2.0$. The risks of energy exhaustion are calculated as the integrated value of $(E_L - E(t))$ (Eq. 9a) and represented as the grey area. **b** *The risks of high glucose levels.* The damages occur when the glucose level exceeds the threshold $x_G^U = 2.0$. The integrated value of the damage $(x_G(t) - x_G^U)$ is calculated as the risks of high glucose level, described as grey area (Eq. 9b)

**Table 1** Fixed parameter values in the calculations

| Description | Symbol | Value |
|---|---|---|
| Efficiency of obtaining glucose from food intake | $a$ | 5.0 |
| Maximum rate of triglyceride production | $a_F$ | 0.5 |
| Minimum rate of triglyceride production | $b_F$ | 0.05 |
| Maximum rate of glycogenesis | $a_L$ | 2.0 |
| Minimum rate of glycogenesis | $b_L$ | 0.05 |
| Maximum rate of glycogenolysis | $a_\beta$ | 0.5 |
| Minimum rate of glycogenolysis | $b_\beta$ | 0 |
| Peak time of glycogenolysis | $\varphi_\beta$ | 18.0 |
| Time period | $\tau$ | 24.0 |
| glucose consumption rate in the active phase | $d_{GA}$ | 1.0 |
| fat consumption rate in the active phase | $d_{FA}$ | 0.5 |
| Relative glucose/fat consumption rate in the resting phase | $\alpha$ | 0.8 |
| Relative ATP production efficiency of fat compared to glucose | $f$ | 3.5 |

### 2.3 *An Index to Measure an Increased Fat Accumulation by Food Intake in the Resting Time*

To study the phase combinations of glycogenesis ($\varphi_L$) and fat synthesis ($\varphi_F$) in which increased fat gain from food intakes in the resting phases is achieved (Salgado-Delgado et al. 2010), we calculated the index for the difference of daily fatty acid accumulation between synchronized and desynchronized food schedule. The index, $\Delta fat$, is given as follows:

$$\Delta fat = \int_{m}^{m+120} \left\{x_F^D(t) - x_F^S(t)\right\} dt, \tag{10}$$

where $x_F^D(t)$ and $x_F^S(t)$ represent the level of fat accumulated at time $t$ under desynchronized and synchronized feeding schedule, respectively. $\Delta fat$ is calculated over 120 h from the starting time $m$. For our calculation, we examined $24 \times 24 = 576$ combinations of $\varphi_F$ and $\varphi_L$ by changing each phase by 1 h.

### 2.4 *Parameter Values Used for Numerical Calculation of the Model*

Parameter values are listed in Table 1. Because we focus on the effects of circadian phases of fat synthesis ($\varphi_F$) and glycogenesis ($\varphi_L$) on glucose metabolism, the phase of glycogenolysis was fixed at $\varphi_\beta = 18$. The peak time of glycogenolysis is fixed at $\varphi_\beta = 18$, the middle of the resting time because rate-limiting enzymes showed peak expression level during the early resting phase (Kim et al. 2017; Greenwell et al. 2019).

Other parameters associated with circadian rhythms were also fixed at constants in Table 1. We fixed $x_G^U = 2$ and $E_L = 2$ in calculations of costs and chose two cycles where start from $n = 84$ in Eqs. (9a) and (9b). We performed the numerical calculation of the differential equations in Eqs. (1), (5), and (6) using "scipy.intergrate.odeint" (Virtanen et al. 2020).

## 3 Results

### 3.1 Minimizing the Risk of Energy Depletion

The risk of energy depletion ($C_E$) was minimized when the timing of fat and glycogen synthesis are desynchronized–the peak time of fat synthesis was set during the resting phase and that of glycogen synthesis was set at the end of the active phase (Fig. 3a; minimum risk was realized at $(\varphi_F, \varphi_L) = (21, 11)$).

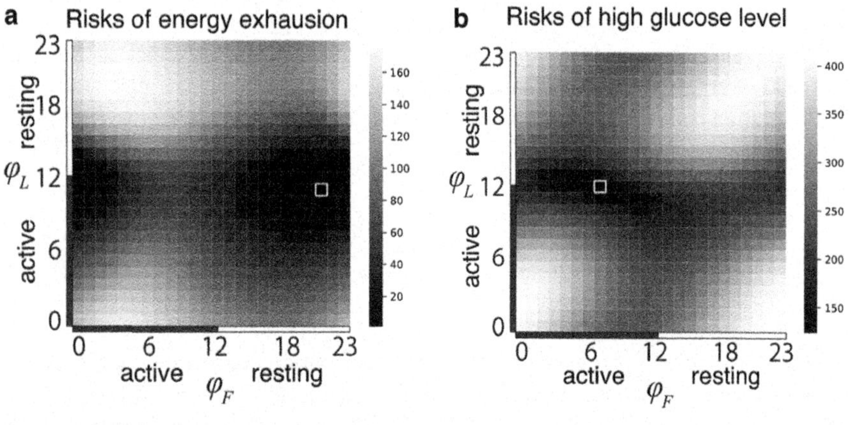


**Fig. 3** *Risks of nutrient homeostasis.* We calculated two possible risks of nutrient homeostasis based on Eq. (9b) and (9c) for $24 \times 24 = 576$ combinations of $(\varphi_F, \varphi_L)$. The light-colored cells represent the higher level of risks, as shown in the scales next to the heat maps. We here search for the condition of $(\varphi_F, \varphi_L)$ that realizes low risk and the condition that achieved the lowest risk is highlighted by the white frame. The black and white bars along the axis represent the active and resting time, respectively. **a** *Risks of energy exhaustion.* The low risks (dark-colored cells) are achieved when $\varphi_F$ is during the resting phase and at the beginning of the active phase and $\varphi_L$ is during the active phase. The lowest risk is at $(\varphi_F, \varphi_L) = (21, 11)$. **b** *Risks of hyperglycemia.* The low risk was achieved when $\varphi_F$ is during the active phase. The lowest risk was achieved at $(\varphi_F, \varphi_L) = (7, 12)$

Minimizing the energy exhaustion during the resting time (i.e. fasting time) is important for energy homeostasis throughout the day. Diel profile for the risk of energy depletion showed the highest risk at the end of resting phase in all combinations of phases for glycogen and fat synthesis (bold lines in Fig. 4a) because of no glucose inputs during the resting phase in the synchronized food schedule (Fig. 1b(I)). The lowest risk of energy depletion was achieved $(\varphi_F, \varphi_L) = (18, 12)$ when fat and glycogen synthesis were peaked at the middle of the resting phase and the end of the active phase, respectively, among the other conditions: $(\varphi_F, \varphi_L) = (6, 0), (6, 12), (18, 0)$; Fig. 4a). The moderate decrease of glucose (solid lines in Fig. 4b) during the resting phase (white area in Fig. 4b) was realized at $\varphi_L = 12$ and constant fat supply (dotted lines in Fig. 4b) throughout the resting phase, which compensates the decline of glucose level, was achieved at $\varphi_F = 18$.

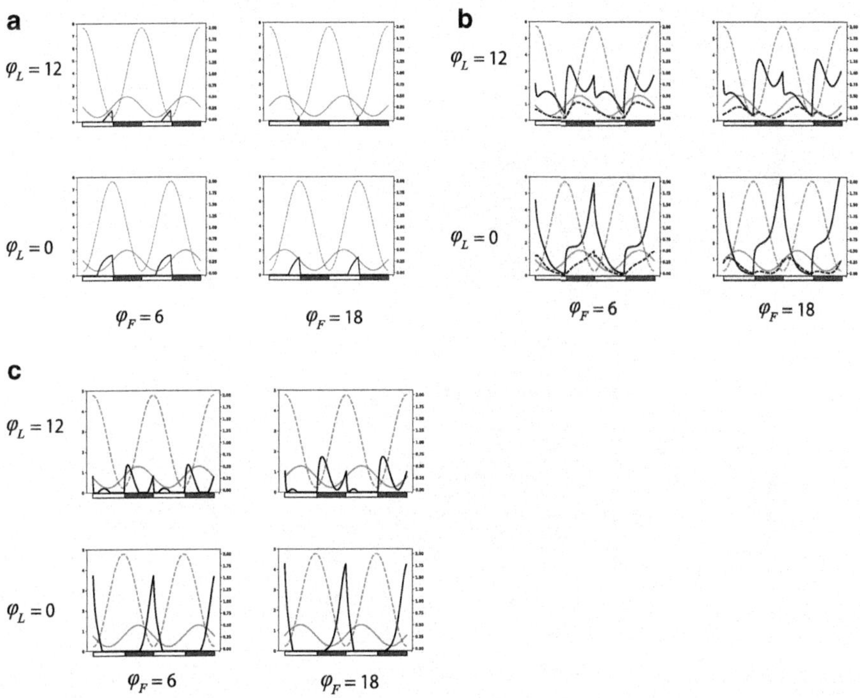


**Fig. 4** *Daily patterns of valuables representing metabolic profiles.* The bold lines show the dynamics of each valuable for two periods for 16 combinations of $(\varphi_F, \varphi_L)$ ($\varphi_F = 6, 18$ and $\varphi_L = 0, 12$). The cosine curves drawn in light-colored bold and dotted lines show the production rate of fat ($\gamma_F$) and glycogen ($\gamma_L$), respectively. The black and white bars at the vertical axis represent the active and resting phases, respectively. **a** *The levels of energy exhaustion.* The bold solid lines show the time series of energy levels below the threshold ($\max\{(E_L - E(t)), 0\}$). **b** *The levels of energy obtained from two resources.* The solid and dotted lines show energy obtained from fat ($d_F x_F$) and glucose ($d_G x_G$), respectively. **c** *The levels of excess glucose.* The solid lines show the level of glucose that exceeds the threshold of damages ($\max\{(x_G(t) - x_G^U), 0\}$)

### 3.2 The Risk of Hyperglycemia is Low When Fat Synthesis is Peaked During the Active Time

The risk of hyperglycemia ($C_G$) was low when fat synthesis is peaked during the active phase (Fig. 3b; minimum risk was realized at $(\varphi_F, \varphi_L) = (7, 12)$). To minimize the excess level of glucose, suppressing the glucose peak during the active phase (i.e. feeding phase) is important. The lowest risk of hyperglycemia was achieved $(\varphi_F, \varphi_L) = (6, 12)$ when fat and glycogen synthesis were peaked at the middle of the active phase and the end of the active phase, respectively, among the other conditions: $(\varphi_F, \varphi_L) = (6, 0), (18, 0), (18, 12)$; Fig. 5c). The higher glucose peak was observed when $\varphi_L = 12$ than ones when $\varphi_L = 0$. If $\varphi_L = 12$, glucose starts to change into glycogen from the beginning of feeding in the active phase and excess glucose can be efficiently stored as glycogen. Moreover, the lower glucose peak was achieved when $\varphi_F = 6$. If the fat production occurs during the active (feeding) phase, the excess glucose can also be changed into fat.

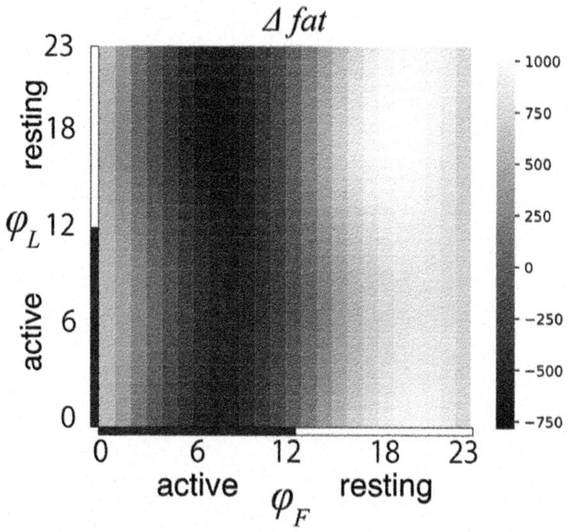


**Fig. 5** *Difference of fat accumulation between food intake in the resting phase and active phase.* $\Delta fat$, the integrated difference of fat between desynchronized and synchronized schedules (Eq. 10) are calculated for each combination of $(\varphi_F, \varphi_L)$. The black and white bars below the axis represent active and resting phases. The scale next to the heat map shows the value of $\Delta fat$. In the heat map, the light-colored cells correspond to the condition of larger $\Delta fat$, which suggests that a desynchronized food schedule is likely to accumulate fat at the condition. The larger $\Delta fat$ are observed when $\varphi_F$ is in the resting phase and the beginning of the active phase

### 3.3 Fat Synthesis Synchronized with Glucose Input Causes Increased Fat Accumulation by Food Intake in the Resting Phase

Increased fat accumulation by food intake in the resting phase ($\Delta fat > 0$; Eq. 10), observed in the previous experimental study (Salgado-Delgado et al. 2010), was achieved when the peak time of fat synthesis ($\varphi_F$) ranged from the resting phase to the beginning of the active phase, regardless of the phase of glycogen synthesis (Fig. 5). This result indicates the importance of synchrony between the fat synthesis peak and the timing of glucose input for the realization of an increased fat accumulation by food intake in the resting phase. The phase of glycogen synthesis has almost no effect on $\Delta fat$.

We here calculated $\Delta fat$ for each of the optimal conditions of $(\varphi_F, \varphi_L)$ that minimized the risks of energy depletion or hyperglycemia. We examined whether $\Delta fat > 0$ holds or not at each $(\varphi_F, \varphi_L)$. At $(\varphi_F, \varphi_L) = (21, 11)$, the optimal condition to minimize the risks of energy depletion, $\Delta fat$ was positive. On the other hand, at $(\varphi_F, \varphi_L) = (8, 12)$, the optimal condition for preventing hyperglycemia, $\Delta fat$ was negative. It suggests that an increased fat accumulation under a desynchronized feeding schedule was realized when the homeostasis of energy was kept. Thus, increased fat accumulation by food intake in the resting phase could be the byproduct of optimization of phases for minimizing the risk of energy depletion. The condition that fat is more likely to be produced during the resting phase ($\varphi_F = 21$) is helpful to avoid energy depletion during the lack of glucose, but simultaneously, it causes excess fat accumulation when food is taken during the resting time. On the contrary, to avoid hyperglycemia, excess glucose needs to be changed into fat during the feeding time, which causes more fat production from food intake during the active phase.

$(\varphi_F, \varphi_L) = (21, 11)$, the phases that achieved both of the minimum risk of energy depletion and $\Delta fat > 0$ in this model, are consistent with the phases of the amount of circulating free fatty acid (Kohsaka et al. 2007) and hepatic glycogen (Doi et al. 2010), respectively, in mice. However, note that $\varphi_F$ and $\varphi_L$ do not represent the phases of the amount of fat and glycogen in this model, but the phases of production rates.

In this model, the time lag between production and release was not considered, and the phases of the molecules and their production rate became overlapped. Doi and colleagues (2010) measured gene expression of *Glycogen synthase 2* (*Gys2*) and it has the opposite diurnal expression pattern to that of hepatic glycogen. Several molecules related to lipid synthesis, such as *fatty acid synthase* (*FAS*) also have the shifted pattern from the free fatty acid (Kohsaka et al. 2007). Considering these molecular processes of production, the dynamics can be modeled precisely.

## 4 Discussion

We here aimed to understand the peripheral circadian regulation of carbohydrate and lipid metabolism in the liver. We developed a mathematical model describing the glucose allocation into glycogen and fat. First, our calculation of costs in terms of energy homeostasis suggested that humans may have a fat production peak during the night to maintain the energy level during the night. Next, we confirmed the condition of the circadian clock in liver that realizes increased fat gain from food intakes in the resting time, which was observed by Salgado-Delgado and colleagues (2010). The situation was realized when the peak time of fat production occurred during the resting phase, especially in the latter part of the resting phase. Our conclusion was not changed by the perturbation of the parameters included in the model $\left(\pm 10\% change in a, a_F, b_F, a_L, b_L, a_\beta, d_{GA}, d_{FA}, \alpha; 0.05 increase in b_\beta\right)$ from the standard values in Table 1. Although our mathematical model is simple, we explained the significance of such metabolic conditions in animals from the point of view of energy homeostasis.

Our model has some room for improvement to describe the more realistic metabolic system. One of the most important elements of the circadian clock is entrainment. Some studies revealed that the peripheral clock in the liver is modulated by food factors (Stokkan et al. 2001; Oike, 2017). To understand the mechanism of entrainment by the meal timings, a multi-scale mathematical model that consists of the gene network and the metabolites dynamics will be useful. Insulin has an important role in the regulation of blood glucose and the entrainment of the clock (Crosby et al. 2019). It will be necessary to consider the effect of insulin on both the glucose intake from the blood into peripheral organs and the entrainment for the prediction of more realistic glucose metabolism.

The pattern of food intake can also be more precisely described if considering the animal model. In this stage, food amount was constant during the daytime in the model. Considering the case in humans, the food intake can be divided into several time points, for example representing breakfast, lunch and dinner. This extension will be useful if the model is used to predict the effect of meal amount balance or timings. Shimizu and colleagues (2018) revealed that the delayed breakfast delayed the clocks in liver and adipose tissues. The mechanism of such phenomena related to meal timings will be explained with the extended model.

Moreover, the length of active and resting phases can be varied to examine the different schedules of activity, which represents the schedule of shift workers. Some epidemiological studies revealed that shift workers are likely to cause obesity (Di Lorenzo et al. 2003). Tsai et al. (2005) shifted the light–dark cycle of rats 12 h three days in a week and revealed that the shifted light exposure enhanced the body weight gain. We can test these situations of shift workers in future work.

The ATP production efficiency of glucose and fat, which are represented by the parameters $f_G$ and $f_F$ (Eq. 7), can be alternative functions for the more detailed description of energy production. We set the relative efficiency of ATP production of fat as $f = f_F/f_G = \frac{106}{30} \cong 3.5$, assuming that fat is 3.5 times more efficient as energy resource in terms of the number of ATPs produced from one molecule. We here ignored the reaction rate of each process. Total efficiency of ATP production can be more precisely described by the number of ATPs that are produced in a certain time period. Thus, $f_G$ and $f_F$ need to be represented as the function of the speed of ATP production and the number of ATPs that can be produced from a certain unit amount of glucose or fatty acids.

Not only in the context of human health, but also the evolution of metabolism, our model has some implications. We found that the fat production peak during the resting phase compensates for the lack of glucose, which implies the benefit of having two independent energy resources. Glucose is used in almost all living organisms as energy resource. Lipid storage is also conserved from bacteria to mammals (Birsoy et al. 2013). The importance of fat accumulation depends on the environment surrounding the living organisms, such as hibernation (Young 1976). It will be possible to represent various styles and conditions of each living organism by changing the metabolic parameters in our model. By calculating the cost function in Chap. 3, we can check the optimal energy resource allocation corresponding to living things in different environments.

**Acknowledgements** This work was done by the JSPS Research Fellowship (DC1) and Grant-in-Aid for JSPS Fellows to A.H (18J20316). We are grateful to Camila Caldana, Yusuke Nakane, and Shigenobu Shibata for helpful advice from the point of view of experimental research. We also thank Shingo Gibo, Hiroshi Ito, Gen Kurosawa, and Motohide Seki for useful comments on mathematical modeling.

## References

Acosta-Rodríguez VA, de Groot MHM, Rijo-Ferreira F, Green CB, Takahashi JS (2017) Mice under caloric restriction self-impose a temporal restriction of food intake as revealed by an automated feeder system. Cell Metab 26(1):267–277.e2. https://doi.org/10.1016/j.cmet.2017.06.007

Adamovich Y, Rousso-Noori L, Zwighaft Z, Neufeld-Cohen A, Golik M, Kraut-Cohen J, Wang M, Han X, Asher G (2014) Circadian clocks and feeding time regulate the oscillations and levels of hepatic triglycerides. Cell Metab 19(2):319–330. https://doi.org/10.1016/j.cmet.2013.12.016

Albers JW, Pop-Busui R (2014) Diabetic neuropathy: mechanisms, emerging treatments, and subtypes. Curr Neurol Neurosci Rep 14(8). https://doi.org/10.1007/s11910-014-0473-5

Antonetti DA, Klein R, Gardner TW (2012) Diabetic retinopathy. N Engl J Med 366(13):1227–1239. https://doi.org/10.1056/NEJMra1005073

Bass J, Takahashi JS (2010) Circadian integration of metabolism and energetics. http://science.sciencemag.org/

Birsoy K, Festuccia WT, Laplante M (2013) A comparative perspective on lipid storage in animals. J Cell Sci 126(7):1541–1552. https://doi.org/10.1242/jcs.104992

Crosby P, Hamnett R, Putker M, Hoyle NP, Reed M, Karam CJ, O'Neill JS (2019) Insulin/IGF-1 drives PERIOD synthesis to entrain circadian rhythms with feeding time. Cell 177(4):896-909.e20. https://doi.org/10.1016/j.cell.2019.02.017

Di Lorenzo L, De Pergola G, Zocchetti C, L'Abbate N, Basso A, Pannacciulli N, Soleo L (2003) Effect of shift work on body mass index: results of a study performed in 319 glucose-tolerant men working in a Southern Italian industry. Int J Obes 27(11):1353–1358. https://doi.org/10.1038/sj.ijo.0802419

Doi R, Oishi K, Ishida N (2010) CLOCK regulates circadian rhythms of hepatic glycogen synthesis through transcriptional activation of Gys2. J Biol Chem 285(29):22114–22121. https://doi.org/10.1074/jbc.M110.110361

Greenwell BJ, Trott AJ, Beytebiere JR, Pao S, Bosley A, Beach E, Finegan P, Hernandez C, Menet JS (2019) Rhythmic food intake drives rhythmic gene expression more potently than the hepatic circadian clock in mice. Cell Rep 27(3):649-657.e5. https://doi.org/10.1016/j.celrep.2019.03.064

Inoki K, Kim J, Guan K-L (2012) AMPK and mTOR in cellular energy homeostasis and grug targets. Annu Rev Pharmacol Toxicol 52(1):381–400. https://doi.org/10.1146/annurev-pharmtox-010611-134537

Kim H, Zheng Z, Walker PD, Kapatos G, Zhang K (2017) CREBH maintains circadian glucose homeostasis by regulating hepatic glycogenolysis and gluconeogenesis. Mol Cell Biol 37(14). https://doi.org/10.1128/mcb.00048-17

Kohsaka A, Laposky AD, Ramsey KM, Estrada C, Joshu C, Kobayashi Y, Turek FW, Bass J (2007) High-fat diet disrupts behavioral and molecular circadian rhythms in mice. Cell Metab 6(5):414–421. https://doi.org/10.1016/j.cmet.2007.09.006

Kumar Jha P, Challet E, Kalsbeek A (2015) Circadian rhythms in glucose and lipid metabolism in nocturnal and diurnal mammals. Mol Cell Endocrinol 418:74–88. https://doi.org/10.1016/j.mce.2015.01.024

Oike H (2017) Modulation of circadian clocks by nutrients and food factors. Biosci Biotechnol Biochem 81(5):863–870. https://doi.org/10.1080/09168451.2017.1281722

Salgado-Delgado R, Angeles-Castellanos M, Saderi N, Buijs RM, Escobar C (2010) Food intake during the normal activity phase prevents obesity and circadian desynchrony in a rat model of night work. Endocrinology 151(3):1019–1029. https://doi.org/10.1210/en.2009-0864

Shan Z, Li Y, Zong G, Guo Y, Li J, Manson JE, Bhupathiraju SN (2018) Rotating night shift work and adherence to unhealthy lifestyle in predicting risk of type 2 diabetes: Results from two large US cohorts of female nurses. BMJ (online) 363:17–19. https://doi.org/10.1136/bmj.k4641

Stokkan KA, Yamazaki S, Tei H, Sakaki Y, Menaker M (2001) Entrainment of the circadian clock in the liver by feeding. Science 291(5503):490–493. https://doi.org/10.1126/science.291.5503.490

Tsai LL, Tsai YC, Hwang K, Huang YW, Tzeng JE (2005) Repeated light-dark shifts speed up body weight gain in male F344 rats. Am J Physiol Endocrinol Metab 289(2 52–2):212–217. https://doi.org/10.1152/ajpendo.00603.2004

Vetter C, Dashti HS, Lane JM, Anderson SG, Schernhammer ES, Rutter MK, Scheer FAJL (2018) Night shift work, genetic risk, and type 2 diabetes in the UK biobank. Diabetes Care 41(4):762–769. https://doi.org/10.2337/dc17-1933

Virtanen P, Gommers R, Oliphant TE et al (2020) SciPy 1.0. fundamental algorithms for scientific computing in Python. Nat Methods 17:261–272. https://doi.org/10.1038/s41592-019-0686-2

Wang XS, Armstrong MEG, Cairns BJ, Key TJ, Travis RC (2011) Shift work and chronic disease: the epidemiological evidence. Occup Med 61(2):78–89. https://doi.org/10.1093/occmed/kqr001

Young RA (1976) Fat, energy and mammalian survival. Integr Comp Biol 16(4):699–710. https://doi.org/10.1093/icb/16.4.699

Zani F, Breasson L, Becattini B, Vukolic A, Montani JP, Albrecht U, Solinas G (2013) PER2 promotes glucose storage to liver glycogen during feeding and acute fasting by inducing Gys2 PTG and GL expression. Mol Metab 2(3):292–305. https://doi.org/10.1016/j.molmet.2013.06.006

# Prediction of Gene Expression Level Using Hierarchical Generalized Linear Model

**Zeyuan Wang, Hong Gu, and Pan Qin**

## 1 Introduction

In eukaryotes, nucleosomes are the basic unit of chromatin, composed of DNA and core histones (Dong and Weng 2013). The structure of core histones is prone to changes, including histone methylation, acetylation, phosphorylation, ubiquitination, etc. These changes are collectively called histone modifications (Peterson and Laniel 2004; Roth et al. 2001), which can affect gene expression mainly through two mechanisms: by changing the spatial structure of DNA, thereby changing the accessibility of transcription factors to DNA; by providing a specific binding surface for the recruitment of transcriptional activators and inhibitors (Li et al. 2007). Unlike DNA mutations, histone modifications are reversible, which makes it meaningful to study the regulatory mechanism of histone modifications on gene expression. Researching on histone modifications can further promote the developing progress of epigenetic drugs for cancer and other diseases (Bannister and Kouzarides 2011; Herceg 2007; Shang 2010).

The research on the regulation of histone modifications on gene expression can be divided into two branches. One is to predict the level of gene expression via developing classification models. A widely accepted method for measuring gene expression level is through the median, which divides the expression values into high level and low level. A high level of expression indicates that the gene is active in the process of transcription and translation, which makes it more likely to be inferred as a cancer driver gene in tumor cells. Therefore, it is of great significance to establish an accurate prediction model. In 2011, Cheng et al. first proposed a method for extracting histone modification signals, and studied both the internal relationship among histone modifications and their influence on the level of gene

Z. Wang · H. Gu · P. Qin (✉)
Faculty of Electronic Information and Electrical Engineering, Dalian University of Technology, 116033 Dalian, China
e-mail: qp112cn@dlut.edu.cn



R. McKibbin et al. (eds.), *Proceedings of the Forum "Math-for-Industry" 2019*, Mathematics for Industry 36, https://doi.org/10.1007/978-981-19-1154-5_8

expression through two-way hierarchical clustering, Support Vector Machine (SVM) and Support Vector Regression (SVR) (Cheng et al. 2011). Dong et al. studied the effects of histone modifications, histone variants, and DNA hypersensitive sites on the level of gene expression through Random Forest in 2012 (Dong et al. 2012). Rutabaga et al. studied the regulation mechanism of five histone modification characteristics on gene expression based on convolutional neural networks in 2016 (Singh et al. 2016). Chitsazian et al. used the Classification and Regression Tree data mining protocol and the Multifactorial Dimensionality Reduction method to predict gene activity based on single histone modification H2BK5ac in 2017 (Chitsazian et al. 2017). However, the median of gene expression is sensitive to data, resulting in the inconsistency of label assignment and a decrease in the classification accuracy. Building a regression model can effectively avoid this problem, which leads to the other branch of research in this field. By implementing a regression model, the degree of impact of different gene loci on gene expression can be further clarified. In 2010, Karlić et al. experimentalized on CD4+T cells and testified that histone modification levels are predictive for gene expression using linear regression (Karlić et al. 2010). Based on this, Costa et al. derived a mixture of linear regression models and took the role of transcription factors into account (Costa and Roider 2011). In 2012, McLeay et al. built log-linear regression-based models that relate gene expression to the binding of 12 different transcription factors, to 7 histone modifications, and to chromatin accessibility (McLeay and Lesluyes 2012). Other regression methods such as regression hidden Markov model (Lee et al. 2014) and SVR model (Cheng et al. 2011; Zhang et al. 2017) have also been studied. However, due to the zero-inflation characteristics of the gene expression data, so far few effective models have been achieved. The SVR model proposed by Cheng et al. reached a correlation coefficient of 0.7 (Cheng et al. 2011). There is still huge room for improvement.

In this research, we took the human B lymphoid (GM12878) cell line as the research object. GM12878 cell line mainly performs the body's humoral immune function with active gene expression, which is suitable for the study. The Transcripts Per Million (TPM) value is used as the standard to measure gene expression (Günter et al. 2012). Since a large part of the genes in the organism is not expressed at the same time, the TPM data contains a large fraction of zeros. Such zero-inflated data are prone to reduce the prediction accuracy of the regression model. To this end, we took the feature of zero inflation into consideration (Lambert 1992) and established a master-slave model under the framework of the generalized linear model (GLM) (Nelder and Wedderburn 1972) consisting of two stages, which first removes the samples whose gene expression values are classified zero by the classification model. After that, the remaining samples go through the second stage, where the regression model is applied to predict the gene expression values.

**Table 1** The locations and functional categories of the ten histone modifications

| Histone modification | Region | Function |
|---|---|---|
| H3K9me3 | Heterochromatin | Inhibition |
| H3K9ac | Enhancer/Promoter | Activation |
| H3K36me3 | Transcription | Structure tag |
| H3K4me1 | Enhancer | Distal action marker |
| H3K4me2 | Transcription | Inhibition |
| H3K4me3 | Promoter | Activation |
| H3K79me2 | Enhancer/Promoter | Inhibition |
| H3K27me3 | Polycomb inhibition | Inhibition |
| H4K20me1 | Transcription | Inhibition |
| H3K27ac | Enhancer/Promoter | Activation |

# 2 Methods

## *2.1 Data Preprocessing*

The histone modification and gene expression data (RNA-seq) used in this paper are from the Encyclopedia of DNA Elements (ENCODE) database (ENCODE 2004). The annotation file of the human reference genome (hg19) comes from the UCSC (http://genome.ucsc.edu/) database, which contains chromosome number, gene name, ENSEMBL-ID, gene type, strand, transcription start sites (TSS), and transcription terminal sites (TTS). Such file covers adequate information of gene location, which plays a vital role in merging gene expression data with histone modification data. In this paper, we study ten histone modification data of the GM12878 cell line. Table 1 lists the ten histone modifications and their located regions and functional categories (Integrative analysis 2015; Grunstein 1997).

The processing of the hg19 annotation file goes as follows. First of all, we retained two sorts of genes, the protein-coding gene, which can guide protein synthesis, and the genes whose expression type is exon. Secondly, for deduplicated IDs, only one of them is kept. We also removed the genes whose annotation information is partially missing, for the reason that they cannot be located on the histone modification data. Finally, 20,114 genes are screened out and generate a 20,114-dimensional response variable $\boldsymbol{y}$ for subsequent analysis.

Since the genes in the RNA-seq data are presented as their ENSEMBL-IDs, which are provided by the ENSEMBL project, in order to convert the ENSEMBL-ID of a gene into its commonly used gene name, an R package called org.Hs.e.g.dg is applied. As the measure of gene expression, the TPM value corresponding to each gene in the RNA-seq file is extracted, denoted by $T_{pm}$. TPM optimized the RPKM method and is a widely adopted method to measure the level of gene expression which can be used to compare gene expression levels in different tissues of the same species. The calculation of TPM is as follows:

$$T_{pm} = \frac{r_g \times r_l \times 10^6}{l_g \times T} \tag{1}$$

$$T = \sum_{g \in G} \frac{r_g \times r_l}{l_g} \tag{2}$$

where $r_l$ denotes the sequencing read length, that is, the average number of nucleotides per sequence read alignment. $l_g$ denotes gene length, and $r_g$ denotes the total number of reads aligned to the corresponding genes.

The original data of ten histone modifications are saved as Bigwig files, downloaded from the ENCODE database. A Bigwig file is a compressed binary file converted from a Wiggle file. Compared with the Wiggle file, it is more space-saving but is not directly usable. For convenience, we use the Wiggle format to represent the histone modification data for its feature of storing real-valued data such as histone modification data and transcriptome data, which can be further implied to extract the characteristic values of histone modifications at different gene loci. The tool bigwigToWig is applied to convert the Bigwig files into Wiggle files with a step length of 25 bp, after which the file is shifted to a wig file with a step length of 50 bp by summing the average value, which is more suitable for the follow-up extraction of gene locus information.

The histone modifications near the gene start sites and termination sites have a significant effect on gene expression, with a feature called long-range effect, that is, the influence sustains over a longer distance. Due to the limitation of gene length, in order to ensure the number of gene samples, we only extracted the histone modification characteristics of 4 kbp before the gene start site and after the gene termination site. The specific extraction method of histone modification data features is shown in Fig. 1. First, each gene is located through the hg19 annotation file. According to the TSS and TTS of each gene, the first 4 kbp gene sequence of each gene TSS is divided into 40 boxes, the 4 kbp sequence after TTS is also divided into 40 boxes; each box is 100 bp in length, and each gene corresponds to a total of 80 boxes. Taking into account the limitation of gene length, the features extracted within the genes are not included. The feature information of the corresponding site is extracted through the histone modification file. The value is extracted every 100 bp, which is used as the feature of the site. For each histone, a total of 40 features before the transcription start site and 40 features after the termination site are extracted. After merging the ten histone modification data, a 20,114×801 design matrix $X$ containing a regular item is created. $\boldsymbol{x}_i \in \mathbb{R}^{801}$ denotes the characteristic value corresponding to certain gene $i$. $y_i$ represents the response variable of the corresponding sample $i$, that is, the overall expression value of gene $i$.

A box-plot of the ten features in the design matrix at each sample is shown in Fig. 2, where one feature is selected for each histone. Due to its wide range of numerical distribution, it is necessary to standardize the feature matrix to guarantee $\boldsymbol{x}_i \in [0, 1]$

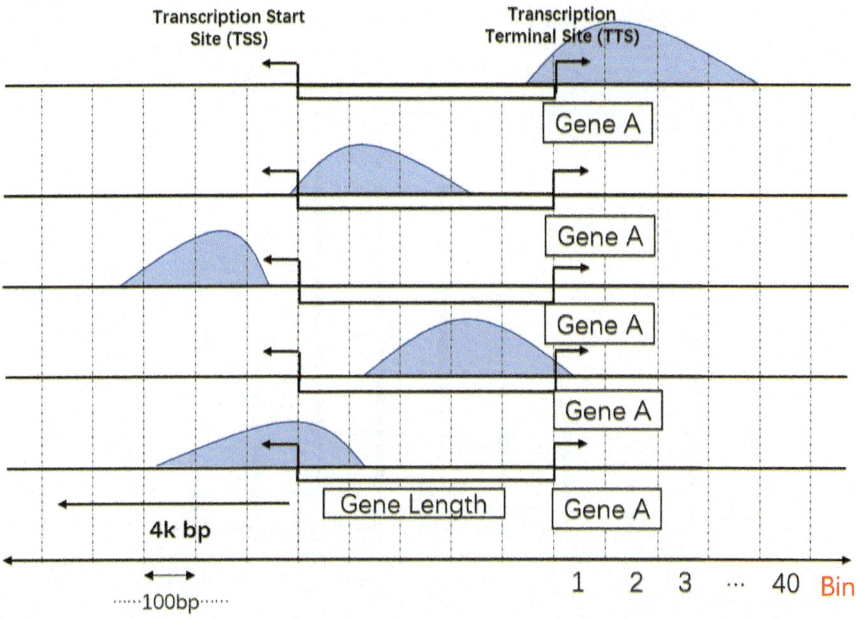


**Fig. 1** The method of histone modified feature extraction

to solve the problem of incomparability between different data and eliminate the adverse effects caused by singular sample data. The specific standardization method is as follows:

$$x_{ij} = \frac{\log_2(x_{ij} + 1)}{\log_2(\max(\boldsymbol{x}_j) + 1)} \tag{3}$$

## 2.2 *Model Establishment*

In this section, the frame of our proposed model is interpreted in detail. In the response variable $\boldsymbol{y}$, zeros take up a large proportion (25.3%). Such a feature is called zero inflation. Building a regression model directly on zero-inflated data will cause severe distortion. For this reason, we propose a master-slave model under the framework of the GLM, dividing the modeling process into two stages. In stage one, a classification model is imposed to distinguish whether the response value is zero. The samples whose response value is not zero enter stage two, where a continuous real-valued regression model based on normal distribution is imposed. The model outputs a list of real numbers as the gene expression values. The detailed explanation of the principle of the master-slave model is described as follows:

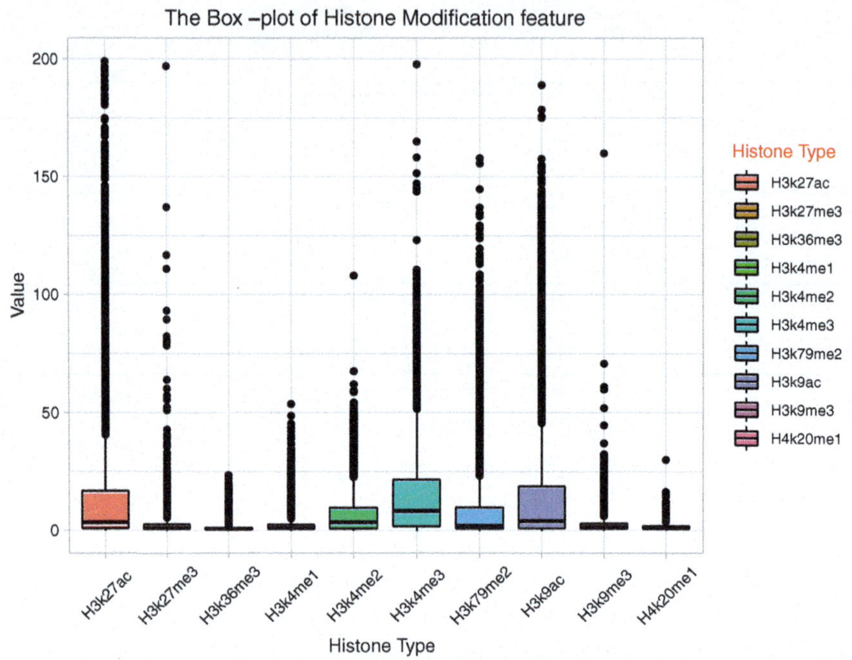


**Fig. 2** The box-plot of histone modification feature before standardization

Stage 1: To implement the classification model, a label for classification is imposed on the response data. The label "0" corresponds to the samples whose response value is zero. Otherwise, the label "1" is assigned. Afterwards, a GLM classification model based on the binomial distribution is used to classify whether the value of the response variable is zero according to the label. According to the given data, for sample $\boldsymbol{x}_i$ of gene $i$, the corresponding response value $y_i$ obeys a binary distribution, whose distribution function is as follows:

$$f_y(y_i|v_i) = P(Y = y_i|v_i) = v_i^{y_i}(1 - v_i)^{1-y_i} \tag{4}$$

where $y_i \in \{0, 1\}, 0 < v_i < 1$, and $E(Y) = v_i$, $Var(Y) = v_i(1 - v_i)$. The logistic function is used as the link function:

$$g(v_i) = \log\left(\frac{v_i}{1 - v_i}\right) = \boldsymbol{x}_i{}^\top \boldsymbol{\beta}_1 \tag{5}$$

where $\boldsymbol{\beta}_1$ is the coefficient vector used for predicting the binary distribution parameter $v_i$. Multiply the distribution functions of all samples to gain the likelihood function:

$$L(\boldsymbol{\beta}_1) = \prod_{i=1}^{n} f_y(y_i|v_i) = \prod_{i=1}^{n} v_i^{y_i}(1 - v_i)^{1-y_i} \tag{6}$$

Take the negative logarithm of the above function to get its negative log-likelihood function, shown as follows:

$$l(\boldsymbol{\beta}_1) = -\log \prod_{i=1}^{n} f_y(y_i|v_i) = -\sum_{i=1}^{n} (y_i \log v_i + (1 - y_i) \log (1 - v_i)) \tag{7}$$

Take the partial derivative of the above formula and set it to zero to get

$$\frac{\partial l}{\partial v_i} = -\frac{y_i}{v_i} + \frac{1 - y_i}{1 - v_i} = 0 \tag{8}$$

where $v_i = \frac{1}{1+\exp(-\boldsymbol{x}_i{}^{\top}\boldsymbol{\beta}_1)}$. By solving the parameter vector $\boldsymbol{\beta}_1$, the classification model can be completely constructed.

Stage 2: The non-zero samples based on the label of the response variable are extracted to construct a regression model. In order to implement the model base on GLM, the response variable is supposed to obey the exponential family distribution. For that purpose, a transformation is imposed on the response variable to make it tend to obey normal distribution. The transformation function is as follows:

$$y_i = \log\left(\frac{y_i + 1}{\text{median}(y_i) + 1}\right) \tag{9}$$

where median $(y_i)$ represents the median of sample $y_i$.

The distribution diagram of the response variable after the transformation is shown as in Fig. 3, through which it can be seen that the values are roughly in the form of a normal distribution. Therefore, a GLM model based on the normal distribution is chosen as the regression model. The distribution function is listed as follows:

$$f(y_i|\mu_i, \sigma_i^2) = \frac{1}{\sqrt{2\pi}\sigma_i} \exp\left(-\frac{(y_i - \mu_i)^2}{2\sigma_i^2}\right) \tag{10}$$

where $\mu_i$ is the mean of the distribution followed by the $i$th sample, and $\sigma_i^2$ is the variance of the distribution. The parameter $\mu_i$ of the normal distribution is connected with the explanatory variable $\boldsymbol{x}_i$ corresponding to the $i$th sample through the link function. The link function is as follows:

$$g(\mu_i) = \mu_i = \boldsymbol{x}_i{}^{\top}\boldsymbol{\beta}_2 \tag{11}$$

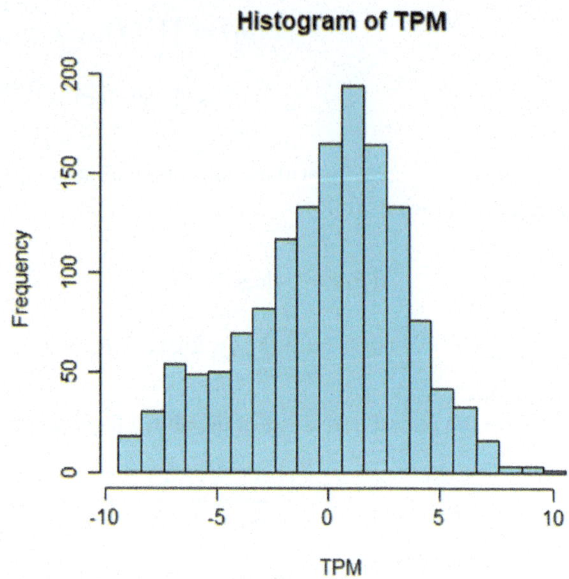


**Fig. 3** The distribution histogram of TPM

where $\boldsymbol{\beta_2}$ is the coefficient vector for predicting the normal distribution parameter $\mu_i$. Similarly, through continuous multiplication, the likelihood function of the expected $\mu_i$ of the GLM based on the normal distribution can be obtained, and $\boldsymbol{\beta_2}$ can be calculated by the maximum likelihood method. The likelihood function form is as follows:

$$L = \prod_{i=1}^{n} \frac{1}{\sqrt{2\pi}\sigma_i} \exp\left(-\frac{(y_i - \boldsymbol{x_i}^\top \boldsymbol{\beta_2})^2}{2\sigma_i^2}\right) \tag{12}$$

and the result is

$$\hat{\boldsymbol{\beta}}_2 = \text{argmax } L \tag{13}$$

In addition to the model mentioned above, the probability density function can be obtained through other techniques such as the Fokker-Planck equation (Risken 1989). The Fokker-Planck equation is one of the most important and powerful tools for studying noise phenomena in systems coupled to a fluctuating environment (Denisov et al. 2009), which can be expressed as follows:

$$\frac{\partial f(y_i, t_i)}{\partial t_i} = \left[-\frac{\partial}{\partial y_i} D_i^{(1)}(y_i) + \frac{\partial^2}{\partial y_i^2} D_i^{(2)}(y_i)\right] f(y_i, t_i) \tag{14}$$

where $t_i$ is called the time coefficient, $D_i^{(2)}(y_i)$ is the diffusion coefficient, and $D_i^{(1)}(y_i)$ is the drift coefficient. The drift and diffusion coefficients may also depend on time. This formulation sheds light on future work and will not be further discussed in this paper.

# 3 Results

In this paper, we establish a regression model to predict gene expression dependent on the characteristics of histone modification. In order to testify the validity of the model, ten sets of histone modification data and gene expression data of the human GM12878 cell line are used. Since the proposed model is a master-slave model, which consists of two stages, different evaluation indicators are used for different stages of the model. The first part of the main model classifies whether the response variable is zero. The classification accuracy and the ROC curve are used to illustrate the performance. In the second part, a regression model acting on the data whose response variable is not classified as zero. The coefficient of determination $R^2$ is calculated, which reflects the performance of the regression model.

The calculation of the classification accuracy is supposed to go through cross-validation, which is a model evaluation method using part of the data each time and repeating the process several times to eliminate contingency. In this article, ten-fold cross-validation is adopted, and the proposed classification model based on GLM reached an accuracy of 0.89, indicating that the model is reliable in distinguishing whether histone modification signal promotes gene expression. Figure 4 shows the ROC curve of the classification model constructed in this paper.

Since most of the current studies on the effect of histone modification data on gene expression levels aim at establishing classification models, relatively few studies made an attempt on predicting gene expression levels by regression methods. In order to testify the effectiveness of the proposed regression model, we made the comparison with some commonly used modeling methods such as multiple linear regression (MLR), SVR, and neural network. Figure 5 is a comparison chart of the fitting degrees of multiple regression algorithms. The result proves that the master-slave model based on GLM outperforms other methods in correlation coefficient and determination coefficient $R^2$, reaching 0.873 and 0.749 separately. The mean square error of the master-slave model is 0.044, indicating the effectiveness of the predicting method. Among the other three regression algorithms, the results of MLR are poor, which means that there is no significant linear relationship between gene expression levels and histone modifications. The performance of the SVR model is average, which shows that the algorithm is not suitable for zero-inflated data. The results of the neural network are relatively better than SVR, which, to a certain extent, proves the feasibility of applying neural networks to this problem. Due to the lack

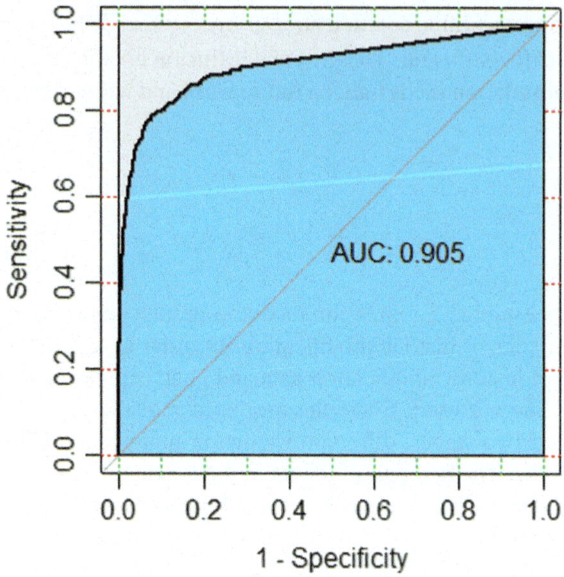


**Fig. 4** The ROC curve of classification model

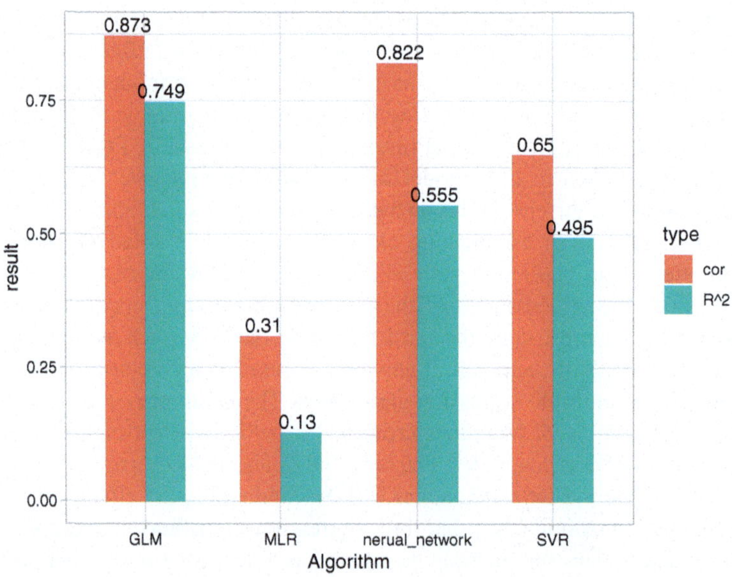


**Fig. 5** Comparison of the fitting degrees of multiple regression algorithms

of interpretability of neural networks, the effect is unpredictable when the structure of the neural network changes. Thus, further research are supposed to be done to explore whether the neural network can reach better performances.

## 4 Conclusion

The mechanism of histone modification affecting gene expression is still not fully understood. However, the strong relationship between them enables us to build a predictive model. This article comprehensively analyzes the human genome-wide annotation data, the histone modification data, and the gene expression data of the GM12878 cell line. Unlike the classification models proposed in previous research, we developed a master-slave regression model based on GLM, which predicts gene expression level dependent on histone modification signals. The master-slave model takes the feature of zero inflation in the response variable into consideration, combining the regression model with a classification model and resulting in an improvement in the accuracy. In future research, other characteristics in the process of gene regulation like the transcription factors are considered to be embedded in our method.

## References

Bannister AJ, Kouzarides T (2011) Regulation of chromatin by histone modifications. Cell Res 21:381–395. https://doi.org/10.1038/cr.2011.22

Cheng C, Yan KK, Yip KY (2011) A statistical framework for modeling gene expression using chromatin features and application to modENCODE datasets. Genome Biol 12. https://doi.org/10.1186/gb-2011-12-2-r15

Chitsazian F, Sadeghi M, Elahi E (2017) Confident gene activity prediction based on single histone modification H2BK5ac in human cell lines. BMC Bioinform 18. https://doi.org/10.1186/s12859-016-1418-6

Costa IG, Roider HG (2011) Predicting gene expression in T cell differentiation from histone modifications and transcription factor binding affinities by linear mixture models. BMC Bioinform 12. https://doi.org/10.1186/1471-2105-12-S1-S29

Denisov SI, Horsthemke W, Hänggi P (2009) Generalized Fokker-Planck equation: derivation and exact solutions. Euro Phys J B 68:567–575. https://doi.org/10.1140/epjb/e2009-00126-3

Dong X, Greven MC, Kundaje A (2012) Modeling gene expression using chromatin features in various cellular contexts. Genome Biol 13. https://doi.org/10.1186/gb-2012-13-9-r53

Dong X, Weng Z (2013) The correlation between histone modifications and gene expression. Epigenomics 5:113–116. https://doi.org/10.2217/epi.13.13

Grunstein M (1997) Histone acetylation in chromatin structure and transcription. Nature 389:349–352. https://doi.org/10.1038/38664

Günter P, Wagner KK, Lynch VJ (2012) Measurement of mRNA abundance using RNA-seq data: RPKM measure is inconsistent among samples. Theory Biosci 131:281–285. https://doi.org/10.1007/s12064-012-0162-3

Herceg Z (2007) Epigenetics and cancer: towards an evaluation of the impact of environmental and dietary factors. Mutagenesis 22:91–103. https://doi.org/10.1093/mutage/gel068

Integrative analysis of 111 reference human epigenomes. Nature 518:317–330 (2015). https://doi.org/10.1038/nature14248

Karlić R, Chung HR, Lasserre J (2010) Histone modification levels are predictive for gene expression. PNAS 107:2926–31. https://doi.org/10.1073/pnas.0909344107

Lambert D (1992) Zero-inflated poisson regression, with an application to defects in manufacturing. Technometrics 34. https://doi.org/10.1080/00401706.1992.10485228

Lee Y, Ghosh D, Zhang Y (2014) Regression hidden Markov modeling reveals heterogeneous gene expression regulation: a case study in mouse embryonic stem cells. BMC Genomics 15:360. https://doi.org/10.1186/1471-2164-15-360

Li B, Carey M, Workman JL (2007) The role of chromatin during transcription. Cell 128. https://doi.org/10.1016/j.cell.2007.01.015

McLeay RC, Lesluyes T (2012) Genome-wide in silico prediction of gene expression. Bioinformatics 28:2789–96. https://doi.org/10.1093/bioinformatics/bts529

Nelder JA, Wedderburn RW (1972) Generalized Linear Models. J R Stat Soc 135:370–384. https://doi.org/10.2307/2344614

Peterson CL, Laniel MA (2004) Histones and histone modifications. Curr Biol 14:546–551. https://doi.org/10.1016/j.cub.2004.07.007

Risken H (1989) The Fokker-Planck equation, 2nd edn. Springer, Berlin

Roth SY, Denu JM, Allis CD (2001) Histone acetyltransferases. Annu Rev Biochem 70:81–120. https://doi.org/10.1146/annurev.biochem.70.1.81

Shang Y (2010) Epigenetics and cancer. In: The Chinese society of biochemistry and molecular biology

Singh R, Lanchantin J, Robins G (2016) DeepChrome: deep-learning for predicting gene expression from histone modifications. Bioinformatics 32. https://doi.org/10.1093/bioinformatics/btw427

The ENCODE (ENCyclopedia Of DNA Elements) project. Science 306:636–640 (2004). https://doi.org/10.1126/science.1105136

Zhang LQ, Li QZ (2017) Estimating the effects of transcription factors binding and histone modifications on gene expression levels in human cells. Oncotarget 8:40090-40103. https://doi.org/10.18632/oncotarget.16988

# Development of Algorithm Improving Label Arrangements in Offset Printing

**Geun Soo Jang, Taehyeong Kim, Hyun-Min Kim, Ki Man Kong, Jeong Rye Park, Jong-Hyeon Seo, Sang-Hyup Seo, and Shin Won Yoon**

**2010 Mathematics Subject Classification** 68U99 · 90C90

## 1 Introduction

A combination with repetition is the number of cases, where $k$ elements are selected from different $n$ elements allowing repetition (Brualdi 2004). It is indicated with the symbol ${}_nH_k$ and the following is established:

G. S. Jang · T. Kim · H.-M. Kim · J. R. Park · S.-H. Seo (✉) · S. W. Yoon
Finance.Fishery.Manufacture Industrial Mathematics Center on Big Data, Pusan National University, Busan 46241, Republic of Korea
e-mail: saibie1677@gmail.com

G. S. Jang
e-mail: sand621@naver.com

T. Kim
e-mail: th_kim@pusan.ac.kr

H.-M. Kim
e-mail: hyunmin@pusan.ac.kr

J. R. Park
e-mail: parkjr@pusan.ac.kr

S. W. Yoon
e-mail: ysw0123@pusan.ac.kr

K. M. Kong
World Komax Co., Ltd., 1505, Centum Jungang-Ro 48, Haeundae-Gu, Busan 48059, Republic of Korea
e-mail: kennethkong@worldkomax.net

J.-H. Seo
Chubu University Academy of Emerging Science, Kasugai 487-0027, Japan
e-mail: hyeonni94@isc.chubu.ac.jp

 
R. McKibbin et al. (eds.), *Proceedings of the Forum "Math-for-Industry" 2019*, Mathematics for Industry 36, https://doi.org/10.1007/978-981-19-1154-5_9

**Fig. 1** Offset printing

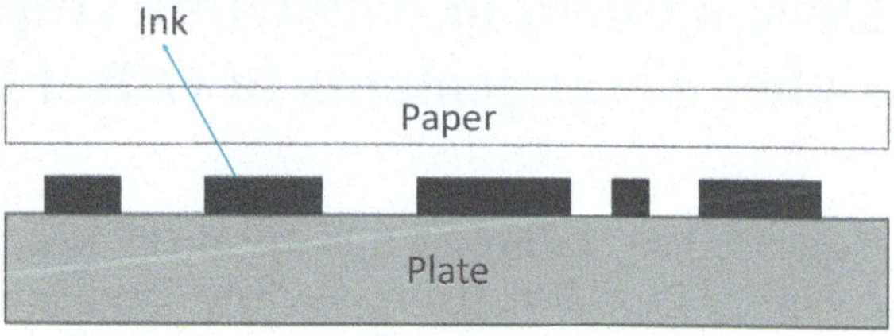


$$_{n}H_{k} = {}_{n+k-1}C_{k} = \frac{(n+k-1)!}{(n-1)!k!}. \quad (1)$$

For instance, the combination with repetition $_2H_4$ to select four elements from among two elements A and B comprises the following five cases.

(1) [A, A, A, A]: a list consisting of four A's;
(2) [A, A, A, B]: a list consisting of three A's and one B;
(3) [A, A, B, B]: a list consisting of two A's and two B's;
(4) [A, B, B, B]: a list consisting of one A and three B's;
(5) [B, B, B, B]: a list consisting of four B's.

Among the above cases, if we want to obtain three A's and nine B's, we can choose [A, B, B, B] × 3. Also, we consider the following as another case.

$$[A, A, B, B] \times 1 + [A, B, B, B] \times 1 + [B, B, B, B] \times 1. \quad (2)$$

In this case, we can get the three A's and nine B's. However, the former case seems to be a 'better' choice because there are three different lists in (2). Let us examine another case.

$$[A, A, A, A] \times 1 + [B, B, B, B] \times 3 - [A] \times 1 - [B] \times 3. \quad (3)$$

[A, B, B, B] × 3 also seems to be a 'better' choice because there is no loss. Under the following conditions, [A, B, B, B] × 3 is the 'best' choice.

(1) Minimize the number of lists.
(2) Minimize the loss of lists.

Offset printing, also called offset lithography, or litho-offset in commercial printing, is a widely used printing technique in which the inked image on a printing plate is printed on a rubber cylinder and then transferred (i.e., offset) to paper or other material. The rubber cylinder gives great flexibility, permitting printing on wood, cloth, metal, leather, and rough paper (see Fig. 1) (Encyclopædia Britannica, inc. 2016).

Offset printing is one of the most common ways of creating printed materials. A few of its common applications include newspapers, magazines, brochures, stationery, and books. Compared to other printing methods, offset printing is best suited

**Fig. 2** Sneaker review : Nike Air Huarache

for economically producing large volumes of high-quality prints in a manner that requires little maintenance (Kipphan 2001). To improve printing process, there were many research in Albert et al. (1979), Musselman (2001), and Carmo (2015). As another improvement, how to make the initial plates is an important issue. The above example shows what is the best arrangement in such a printing method.

In the past, production was based on the ordering of products from companies and predictions of consumption. However, as the Internet market has become popular, the production systems have been altered by consumers. Now, many factories produce only products ordered by consumers.

World Komax is a company that produces labels using offset printing. The labels refer to stickers containing bar-codes attached to garments or shoes as follows (see Fig. 2). Each bar-code in the label contains fixed information such as product names and colors, and variable information such as the manufacturing date.

Since the program of World Komax is not suitable for small quantity batch productions, the number of label losses increased compared to the past.

Section 2 of this paper will describe the process of label printing using offsets. The modeling of the problem will be carried out in Sect. 3, and examples to help the understanding of the problem will be prepared in Sect. 4. The final results of the algorithm will be described in Sect. 5.

## 2 Offset Label Printing Process

### *2.1 Label Printing Process*

Before discussing the process of label printing, we define the following terms.

* **Plate**: A printing plate for the offset printing (see Fig. 3).
* **Loss**: The number of labels printed in excess of the order-quantity.

The offset label printing process is as follows. First, we receive orders from customers. The order includes many types of labels and order-quantities by type (see Fig. 4). Thereafter, offset printing plates are made. Many types of labels are placed

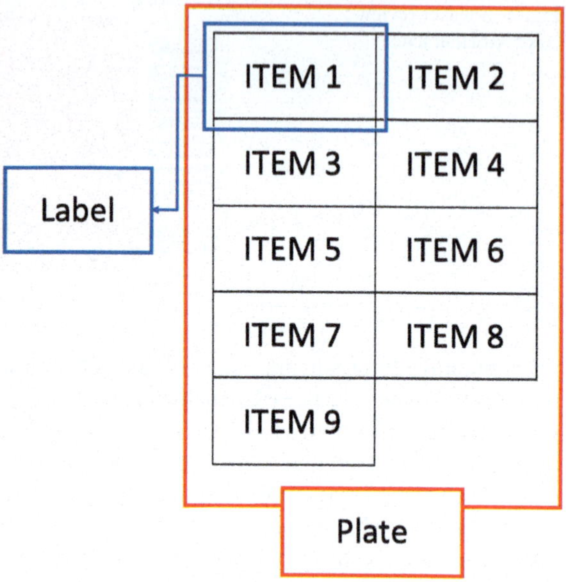


**Fig. 3** Plate and label

on each plate so that many labels are printed at once. After the plates are made, we produce more labels than order-quantities using each plate. Then, the sheets are cut into the label size. For the final process, the labels are collected by type.

## 2.2 *Major Points for Cost Saving*

The constraints and major points that will be considered in this paper for cost-saving are as follows. First, one type of label should be placed on only one plate. This is to prevent different types of labels from being mixed when collected by type after the printed sheets are cut. Meanwhile, the total number of labels placed on each plate is also constant because the sizes of individual plates are constant and the sizes of labels in one order are also constant. In addition, the number of plates should be as minimized as possible because plates are made using molds and the costs are high. Finally, the Loss should be minimized because overprinted labels cannot be used and should be entirely discarded.

## 2.3 *Sorting Reports Output Program*

The production of plates and the use of printing paper incur costs. To reduce the costs, order details are input to output appropriate methods to place labels on the plate as

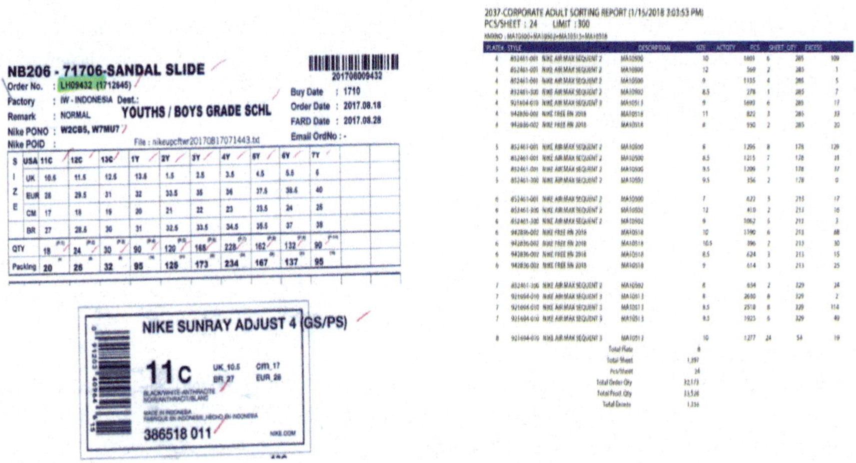


**Fig. 4** An order form (left) and a sorting report (right). *Source* World Komax Co., Ltd.

sorting reports. The plate makers produce plates according to the instructions in the sorting reports (see Fig. 4).

Since the existing sorting report output method was not suitable for small quantity batch production systems, the algorithm had to be improved. Therefore, this study was conducted to develop new algorithms suitable for small quantity batch production systems too.

## 3 Modeling and Flowchart

To formulate our problem as a mathematical optimization problem, we first need to define our notation.

- Let $I$ be a set of products.
- For each product $i \in I$, let $b_i$ be the number of orders.
- Let $\mathbf{b} = (b_i | i \in I)$ be a vector of order-quantities.
- $k$ is the total number of labels that can be placed in one Plate.
- $\pi$ is a partition of $I$ such that $1 \leq |P| \leq k$ for any $P \in \pi$.
- Let $\Gamma_\pi$ be a set of matrix $A \in \mathrm{Mat}_{\pi \times I}(\mathbf{Z})$ satisfying the following:
  - For all $(P, i) \in \pi \times I$, $A_{P,i} \geq 0$, and $A_{P,i} = 0$ if and only if $i \notin P$.
  - For each $P \in \pi$, $\sum_{i \in P} A_{P,i} = k$.

Using the above notation,

$$\left\lceil \max \left\{ \frac{b_i}{A_{P,i}} \middle| i \in P \right\} \right\rceil \tag{4}$$

is the printing number of Plate $P$, where $\lceil\ \rceil$ means the ceiling. Assume that $\alpha$ is the cost to produce one Plate, and $\beta$ is the cost of the loss of one label. Then, our goal is to obtain the following:

$$\min_{\pi}\{\alpha|\pi| + \beta E_{A,b} | A \in \Gamma_\pi\} \tag{5}$$

where

$$E_{A,b} = \sum_{P\in\pi}\sum_{i\in I}\left(\left\lceil \max\left\{\frac{b_i}{A_{P,i}}\middle| i \in P\right\}\right\rceil \cdot A_{P,i} - b_i\right) \tag{6}$$

means the total number of losses of labels.

Naturally, the complete enumeration using combinations with repetition is the surest way. However, this method takes too much time. For instance, when $n = 65$ and $k = 24$, the combination with repetition ${}_{65}H_{24}$ comprises about $2.36 \times 10^{21}$ cases. Then, the calculation takes more than 658 h, that is, more than 27 days using a super computer that can calculate $10^{15}$ partitions per second. Given that there are time constraints from the date of order receipts to the delivery date, this is a very long computation time.

In this algorithm, this problem was solved by introducing any positive integers as thresholds. Thresholds mean the allowed amount of losses occurring in each plate. Adopting a partition that does not exceed the threshold will dramatically reduce the time taken. Based on the foregoing, the flowchart of the algorithm can be set forth as follows (see Fig. 5).

In Fig. 5, $z$ is the threshold and the others are explained in the notation above. This algorithm outputs matrix $A = \bigoplus_P A_{P,i}$ containing the label of each product when $I$ and $\mathbf{b}$ have been input for $z, k, num$. $P$, which is one of the results of $loop(k, z, num)$, is the set of products which are contained in the Plate (see Fig. 6). By removing $P$ from $I$, one type of label should be placed on only one plate. The algorithm repeats until $I$ is empty.

Meanwhile, in the case of the $loop(k, z, num)$ function, the flowchart in Fig. 6 should be followed. First, the $Part(k, num)$ function finds partitions using the combination with repetition ${}_{num}H_k$ which is indicated in the form of a list. We set the result as *Part_list*. For example, let $k = 6$ and $num = 3$, then $Part\_list = Part(6, 3) = \{[4, 1, 1], [3, 2, 1], [2, 2, 2]\}$. $N$ is a printing number that is expressed in (4). An appropriate $P$ that has $num$ pieces of products is selected from $I$ and the loss is obtained using *Part_list* and the printing number.

If the loss exceeds the threshold $z$, another *Part_list* will be selected and the foregoing will be repeated while adjusting $num$ until the threshold $z$ is not exceeded. As a result of this process, $A_{P,i}$ whose loss does not exceed the threshold is obtained.

**Fig. 5** The main flowchart

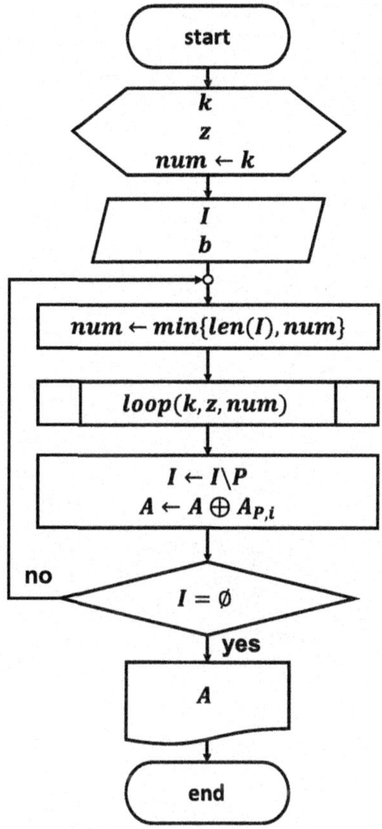


## 4 Examples

This example was described to help the understanding of the problem. For the next two examples, we assume that $k$ is equal to 4.

**Example 1** Assume that $I = \{1, 2, 3\}$, and $\mathbf{b} = (50, 30, 20)$.

We consider $\pi = \{\{1, 2\}, \{3\}\}$, a partition that matches $\mathbf{b}$. Without loss of generality, assume that $P_1 = \{1, 2\}$, $P_2 = \{3\}$. Since $k = 4$, the matrix $A$ can be found as follows:

$$A = \begin{pmatrix} 2 & 2 & 0 \\ 0 & 0 & 4 \end{pmatrix}. \tag{7}$$

In this case, the printing numbers (4) of $P_1$ and $P_2$ are as follows, respectively:

$$\left\lceil \max \left\{ \frac{b_i}{A_{P_1,i}} \middle| i \in P_1 \right\} \right\rceil = \left\lceil \max \left\{ \frac{50}{2}, \frac{30}{2} \right\} \right\rceil = 25 \tag{8}$$

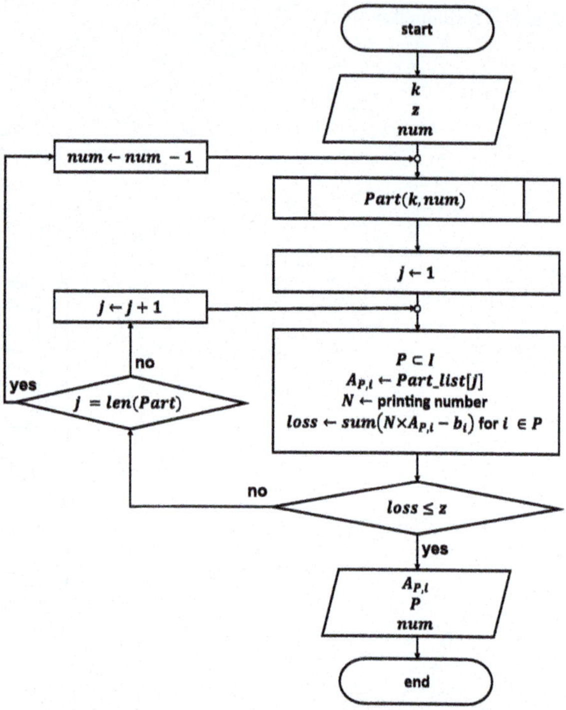


**Fig. 6** The Flowchart of $loop(k, z, num)$

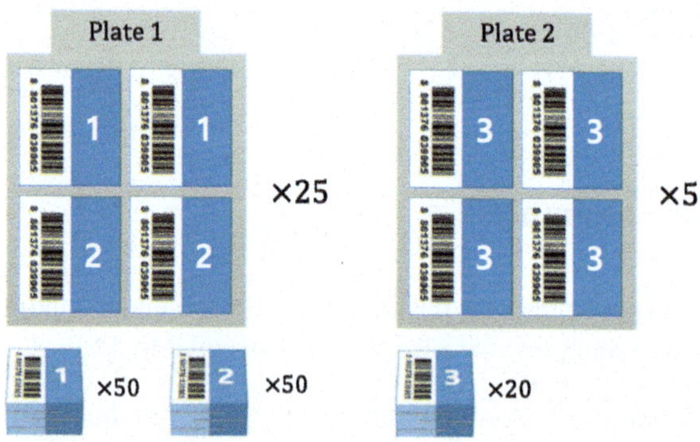


**Fig. 7** Visualization of Example 1 for $\pi = \{\{1, 2\}, \{3\}\}$

and

$$\left\lceil \max \left\{ \frac{b_i}{A_{P_2,i}} \middle| i \in P_2 \right\} \right\rceil = \left\lceil \max \left\{ \frac{20}{4} \right\} \right\rceil = 5. \tag{9}$$

We can see Fig. 7 for more details.

It can be seen that the total number of Loss $E_{A,\mathbf{b}}$ is 20.

Now, we consider a new partition $\pi = \{\{1, 2, 3\}\}$. In this case, $A = (2\ 1\ 1)$, and printing number (4) is

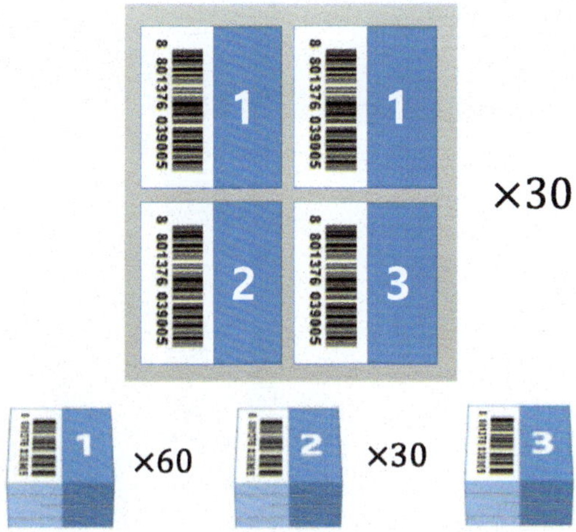


**Fig. 8** Visualization of Example 1 for $\pi = \{\{1, 2, 3\}\}$

$$\left\lceil \max\left\{ \frac{b_i}{A_{P,i}} \middle| i \in P \right\} \right\rceil = \left\lceil \max\left\{ \frac{50}{2}, \frac{30}{1}, \frac{20}{1} \right\} \right\rceil = 30. \tag{10}$$

In addition, it can be easily seen that the total number of Loss $E_{A,\mathbf{b}}$ is 20 (see Fig. 8). The array shown in Fig. 8 is more efficient since its Loss is the same but its number of plates is smaller.

**Example 2** Assume that $I = \{1, 2\}$, and $\mathbf{b} = (50, 20)$.

We consider $\pi = \{\{1, 2\}\}$, a partition that matches $\mathbf{b}$. Since $k = 4$, the matrix $A = (2\ 2)$, and the printing number (4) is

$$\left\lceil \max\left\{ \frac{b_i}{A_{P,i}} \middle| i \in P \right\} \right\rceil = \left\lceil \max\left\{ \frac{50}{2}, \frac{20}{2} \right\} \right\rceil = 25. \tag{11}$$

The total number of Loss $E_{A,\mathbf{b}}$ is 30 (see Fig. 9).

For the same partition $\pi$, matrix $A = (3\ 1)$ can be considered. In this case, the printing number (4) is

$$\left\lceil \max\left\{ \frac{b_i}{A_{P,i}} \middle| i \in P \right\} \right\rceil = \left\lceil \max\left\{ \frac{50}{3}, \frac{20}{1} \right\} \right\rceil = 20 \tag{12}$$

and the total number of Loss $E_{A,\mathbf{b}}$ is 10 (see Fig. 10). The array shown in Fig. 10 is more efficient because its number of plates is the same but fewer losses occur.

If there are too many products and large order-quantities, it requires a lot of iteration. Now, we consider a real problem of World Komax. In this case, $k = 18$, $\alpha = 300$, and $\beta = 1$. Table 1 is an actual sorting report of an order from World Komax.

As we can see from Table 1, there are 14 products and the largest order-quantity is 1059. It uses 3 plates and total Loss is 611. Total cost is 1511.

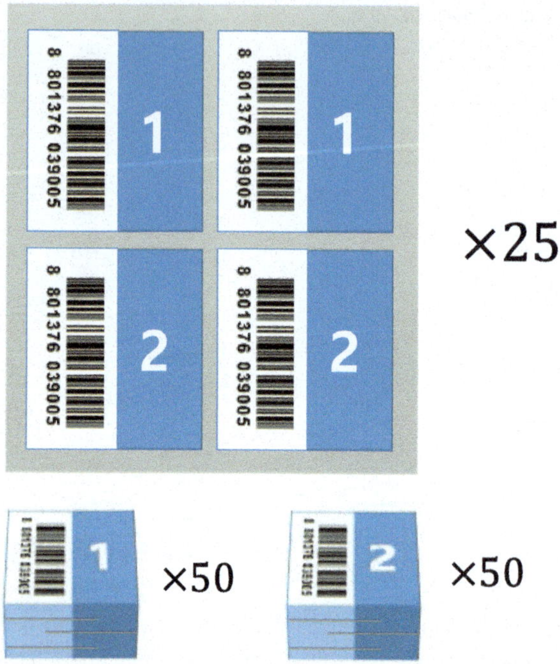


**Fig. 9** Visualization of Example 2 for $\pi = \{\{1, 2\}\}$ and $A = (\,2 \quad 2\,)$

**Table 1** The sorting report of World Komax

| Plate | Product | Order-quantity | psc | Printing number | Production | Loss |
|---|---|---|---|---|---|---|
| 1 | 1 | 59 | 2 | 32 | 64 | 5 |
| | 2 | 156 | 5 | | 160 | 4 |
| | 3 | 9 | 1 | | 32 | 23 |
| | 4 | 162 | 6 | | 192 | 30 |
| | 5 | 102 | 4 | | 128 | 26 |
| 2 | 6 | 1059 | 5 | 228 | 1140 | 81 |
| | 7 | 886 | 4 | | 912 | 26 |
| | 8 | 228 | 1 | | 228 | 0 |
| | 9 | 832 | 4 | | 912 | 80 |
| | 10 | 862 | 4 | | 912 | 50 |
| 3 | 11 | 532 | 4 | 133 | 532 | 0 |
| | 12 | 532 | 5 | | 665 | 133 |
| | 13 | 482 | 4 | | 532 | 50 |
| | 14 | 582 | 5 | | 665 | 103 |

**Table 2** The result of our algorithm

| Plate | Product | Order-quantity | psc | Printing number | Production | Loss |
|---|---|---|---|---|---|---|
| 1 | 7 | 886 | 5 | 178 | 890 | 4 |
| | 10 | 862 | 5 | | 890 | 28 |
| | 11 | 532 | 3 | | 534 | 2 |
| | 12 | 532 | 3 | | 534 | 2 |
| | 4 | 162 | 1 | | 178 | 16 |
| | 2 | 156 | 1 | | 178 | 22 |
| 2 | 6 | 1059 | 9 | 118 | 1062 | 3 |
| | 14 | 562 | 5 | | 590 | 28 |
| | 8 | 228 | 2 | | 236 | 8 |
| | 5 | 102 | 1 | | 118 | 16 |
| | 1 | 59 | 1 | | 118 | 59 |
| 3 | 9 | 832 | 11 | 76 | 836 | 4 |
| | 13 | 482 | 7 | | 532 | 50 |
| 4 | 3 | 9 | 18 | 1 | 18 | 9 |

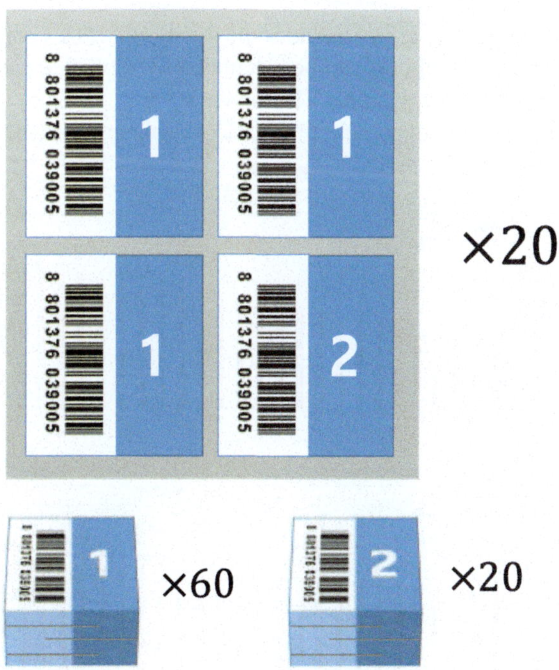


**Fig. 10** Visualization of Example 2 for $\pi = \{\{1, 2\}\}$ and $A = (\ 3 \quad 1\ )$

**Example 3** We obtain another partition of the sorting report using our algorithms. Table 2 shows this result.
The matrix $A$ can be found as follows.

$$\begin{array}{c} \text{column index} \\ \begin{array}{cccccccccccccc} 7 & 10 & 11 & 12 & 4 & 2 & 6 & 14 & 8 & 5 & 1 & 9 & 13 & 3 \end{array} \end{array}$$

$$A = \begin{pmatrix} 5 & 5 & 3 & 3 & 1 & 1 & 0 & 0 & 0 & 0 & 0 & 0 & 0 & 0 \\ 0 & 0 & 0 & 0 & 0 & 0 & 9 & 5 & 2 & 1 & 1 & 0 & 0 & 0 \\ 0 & 0 & 0 & 0 & 0 & 0 & 0 & 0 & 0 & 0 & 0 & 11 & 7 & 0 \\ 0 & 0 & 0 & 0 & 0 & 0 & 0 & 0 & 0 & 0 & 0 & 0 & 0 & 18 \end{pmatrix}$$

The total Loss $E_{A,b}$ is 251. So, total cost is $\alpha|\pi| + \beta E_{A,b} = 300 \times 5 + 1 \times 251 = 1451$. We use one more Plate, but the total cost is reduced by 60.

## 5 Result

Each sorting report includes the number of losses corresponding to one plate. Using this, the total cost can be calculated by replacing each plate with loss. We used 82 sorting report samples. The total cost was reduced by a minimum of −6.85% (sample no. 15) to a maximum of 27.5% (sample no. 74) (see Fig. 11).

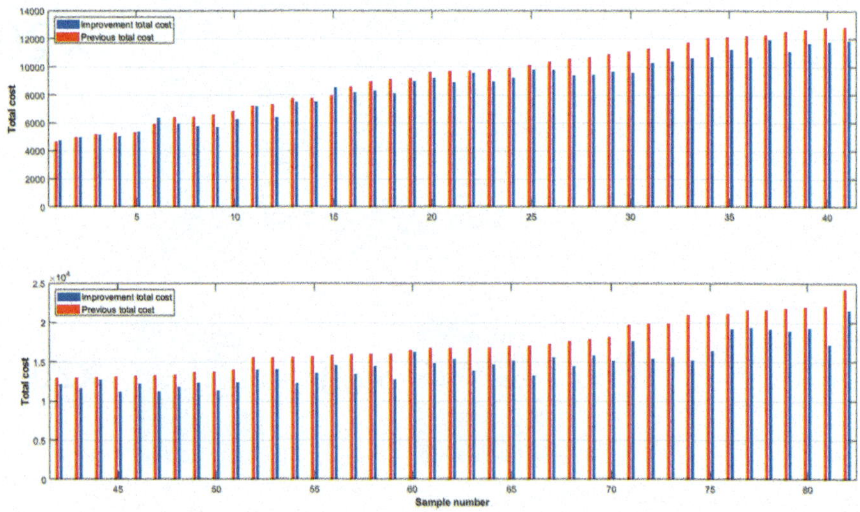


**Fig. 11** Comparing the Results

We used the *paired t-test* (Rice 2006) to verify the efficiency of the algorithm. The paired t-test is one of the two-sample t-tests, and it is a test that verifies whether the two groups are different. The two populations are as follows.

- population1: total cost before applying the algorithm;
- population2: total cost after applying the algorithm;
- sample1: sample of 82 items from population1;
- sample2: sample of 82 items from population2.

In order to proceed with the two-sample t-test, the two groups have to satisfy normality and homoscedasticity. We choose 82 sample data from the two populations, which can satisfy normality by the *central limit theorem* (Durrett 2004). In addition, we identified the homoscedasticity of the two samples through the *var.test* of R. R is a programming language and free software environment for statistical computing and graphics.

The null hypothesis ($H_0$) of var.test is that 'the variances of the two groups are equal', and the alternative hypothesis ($H_1$) is that 'the variances of the two groups are different'. If the $p$-value is below the significance level, the null hypothesis ($H_0$) will be rejected and if the $p$-value is not lower than the significance level, the alternative hypothesis ($H_1$) will be rejected. The result of var.test with the significance level of 0.05 is as follows (see Fig. 12).

Since the $p$-value is not lower than the significance level (0.05), the alternative hypothesis ($H_1$) is rejected and the null hypothesis ($H_0$) is accepted. We say that the variances of the two groups can be equal.

Since the two groups satisfy normality and homoscedasticity, we verified whether the difference between the two groups is significant through paired t-test. In this test, the null hypothesis ($H_0$) is 'the total cost will be the same after applying the algorithm,' and the alternative hypothesis ($H_1$) is 'the total cost will be reduced after applying the algorithm.' The result of the paired t-test with the significance level of 0.05 is as follows (see Fig. 13).

Since the $p$-value is below the significance level (0.05), the null hypothesis ($H_0$) is rejected and the alternative hypothesis ($H_1$) is accepted. We say that the difference between the two groups can be significant.

```
> var.test(sample1,sample2)

        F test to compare two variances

data:  sample1 and sample2
F = 1.5552, num df = 81, denom df = 81, p-value = 0.04847
alternative hypothesis: true ratio of variances is not equal to 1
95 percent confidence interval:
 1.002998 2.411501
sample estimates:
ratio of variances
          1.555227
```

**Fig. 12** var.test of R

```
> t.test(sample1,sample2, paired=TRUE)

        Paired t-test

data:  sample1 and sample2
t = 10.766, df = 81, p-value < 2.2e-16
alternative hypothesis: true difference in means is not equal to 0
95 percent confidence interval:
 1240.884 1803.531
sample estimates:
mean of the differences
               1522.207
```

**Fig. 13** t.test of R

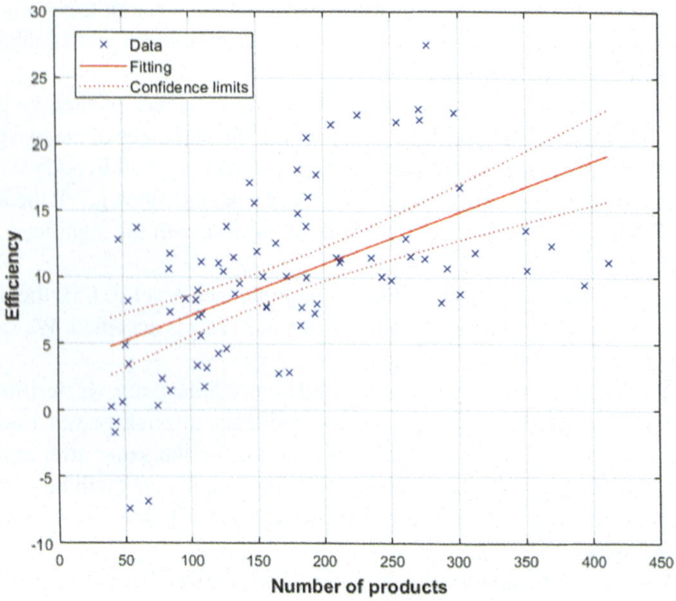


**Fig. 14** The linear fitting for the efficiency over the number of products

Finally, we check the efficiency over the number of products which is the size of $I$ (cf. Sect. 3). We see that the efficiency increases as the number of products increases in Fig. 14. For each of the 82 samples, we calculated the efficiency as the following formula.

$$\text{Efficiency} = \left[1 - \frac{\text{Improvement Total Cost}}{\text{Previous Total Cost}}\right] \times 100. \tag{13}$$

The result of the linear fitting shows that the efficiency and the number of products have a positive correlation. Thus, we see that the improved algorithms have better results in many products. This suggests that the algorithm fits well into small quantity batch productions.

**Notice**. Please note that the detailed idea of the algorithm cannot be described for company confidentiality reasons.

**Acknowledgements** This work was supported by the National Research Foundation of Korea (NRF) grant funded by the Korean Government (MSIP) (NRF-2017R1A5A1015722, NRF-2018R1D1A1B07048197). The authors extend thanks to Prof. Sang-il Kim of Pusan National University who provided much support and advice for writing this paper.

## References

Brualdi RA (2004) Introductory combinatorics. Pearson Prentice Hall

Chan AW, Udo Graefe PW (1979) An interactive sequencing aid for printing presses. Comput Indus Eng 3(4):321–325

do Carmo AB, Raeder M, Nunes T, Kolberg M, Fernandes LG (2015) A job profile oriented scheduling architecture for improving the throughput of industrial printing environments. Comput Indus Eng 88:191–205

Durrett R (2004) Probability: theory and examples, 4th edn. Cambridge University Press

Encyclopædia Britannica, inc. "offset printing" (2016). https://www.britannica.com/technology/offset-printing

Kipphan H (2001) Handbook of print media: technologies and production methods. Springer, Berlin

Musselman KJ (2001) Complex scheduling of a printing process. Comput Indus Eng 39(3–4):273–291

Rice JA (2006) Mathematical statistics and data analysis, 3rd edn. Duxbury Advanced

The manufacturer's authorised representative in the EU is Springer Nature Customer Service Centre GmbH, Europaplatz 3, 69115 Heidelberg, Germany. If you have any concerns regarding our products, please contact ProductSafety@springernature.com

Printed and bound by CPI Group (UK) Ltd, Croydon, CR0 4YY
30/09/2026
02225336-0002